PLANNING, DESIGN, AND IMPLEMENTATION FOR NEW AND UPGRADED WATER RESOURCE RECOVERY FACILITIES

WEF Manual of Practice No. 28

Second Edition

2021

Water Environment Federation
601 Wythe Street
Alexandria, VA 22314–1994 USA
http://www.wef.org

About WEF

The Water Environment Federation (WEF) is a not-for-profit technical and educational organization of 35,000 individual members and 75 affiliated Member Associations representing water quality professionals around the world. Since 1928, WEF and its members have protected public health and the environment. As a global water sector leader, our mission is to connect water professionals; enrich the expertise of water professionals; increase the awareness of the impact and value of water; and provide a platform for water sector innovation. To learn more, visit www.wef.org.

Prepared by the **Planning, Design, and Implementation for New and Upgraded Water Resource Recovery Facilities** Task Force of the Water Environment Federation

José Velazquez, PE, BCEE,
    Chair

Kyle J. Blunn
Marie S. Burbano, PhD, PE,
    BCEE
Keli Callahan, PE
Chein-Chi Chang
James Chelius, PE
Rebecca Dahdah, PE
Richard Finger
Georgine Grissop, PE, BCEE
Jenny Hartfelder, PE
Brian S. Jessee, PE
Murthy Kasi
EJ Katsoulas
Jing Liao

Arlene M. Little, PE
Yanjin Liu, PhD, PE
Jose Christiano Machado Jr.,
    PhD, PE
Jason Maskaly
Jane McLamarrah
Arifur Rahman, MRRDC
    Publications Subcommittee
John Scheri, PE, BCEE
Amber Schrum
Norbert Schulz
Brian Shell
Nicole Stephens, PE
Tanush Wadhawan
Jianfeng Wen, PhD, PE
Tina Wolff, PE
Kameryn Wright, PE

Under the direction of the **Municipal Resource Recovery Design Subcommittee of the Technical Practice Committee**

Manuals of Practice of the Water Environment Federation

The WEF Technical Practice Committee (formerly the Committee on Sewage and Industrial Wastes Practice of the Federation of Sewage and Industrial Wastes Associations) was created by the Federation Board of Control on October 11, 1941. The primary function of the Committee is to originate and produce, through appropriate subcommittees, special publications dealing with technical aspects of the broad interests of the Federation. These publications are intended to provide background information through a review of technical practices and detailed procedures that research and experience have shown to be functional and practical.

Water Environment Federation Technical Practice Committee Control Group

Contents

List of Figures

List of Tables

Preface

Many municipalities and similar governmental entities are faced with making decisions on how to provide residents with cost-effective, environmentally-sound methods to manage wastewater, protect the water environment, and use resources wisely. These challenges may take the form of upgrading from individual or non-centralized systems to new regional mechanical facilities or upgrading their water resource recovery facility with new and higher capacity systems. Owners and managers are faced with the task of planning, coordinating, financing, and managing these upgrades. Because these often occur once every 20 to 30 years at small to medium sized facilities, most owners have little to no experience. This manual serves as a guidance document to help utility managers and others, such as city engineers, public works directors, regulators, and contractors, through an upgrade. It includes sections on how to procure an engineer, how do develop a scope of service for the facility plan, design and construction services, how to evaluate engineering proposals, how to make an informed decision on how to move to more complex facilities, how to evaluate and select different project delivery methods, how to interact with the engineer during each phase, and other topics. In addition, it covers what can be expected during construction, what to expect during startup, and simply how to survive this process.

This second edition of the manual was produced under the direction of José Velazquez.

Authors' and reviewers' efforts were supported by the following organizations:

Advanced Engineering and Environmental Services
Butler, Fairman & Seufert, Inc.
Carollo Engineers, Inc.
Kokosing Industrial
Lamp Rynearson
Mott MacDonald, LLC
Stantec Consulting Services, Inc.
Tetra Tech

1

Introduction

José Velazquez, PE, BCEE

1.0 PURPOSE OF MANUAL

The first edition of the Manual of Practice No. 28 was published in 2005. This second edition has been prepared to update the concepts and ideas presented in the original manual. Significant updates to the discussion on project implementation have been made. The overall organization of the manual has been revised to better reflect current trends in planning, design, and implementation of water reclamation projects.

Most major water treatment facility upgrades occur infrequently, sometime every 20 to 25 years or more. Therefore, many owners (and facility or utility managers) experience only one in their careers. In some cases, utility providers are faced with upgrading from localized, individual systems to more centralized (and typically more complex) regional ones. Going through a major upgrade, from facility planning to startup, is a complex process.

Owners are faced with questions such as:

* Why is the project needed?
* When is the project needed?
* What are potential funding sources?
* What is the best way to procure engineering services?
* What needs to be included in the scope of services?

- What is the best way to deliver the final project?
- How does one ensure the most economical and successful project?
- What equipment is best to procure, operate, and maintain?
- How does one keep the facility running during construction?
- How does one ensure that employees are trained to operate the new facility?

The concept for this manual is to take the experience of owners who have been through a major upgrade from start to finish as well as the expertise of engineers, construction contractors, and construction managers who have been involved with many upgrades, and make their experiences available so that future projects are managed efficiently, economically, and successfully. These three major entities—owners, engineers, and contractors—make up the team that is needed for a successful project, so their contributions to this manual are invaluable.

The primary purpose of this manual is to ensure that facility projects, both new and upgrades, are completed safely; facilities continue to protect public health and the environment during construction; water and other resources are recovered for reuse; and good design and construction result in high-quality facilities that can be easily, properly, and economically operated from the first day of operations.

The goal of this manual is to serve as a guidance document for owners, city and town engineers, public works directors, regulators, consulting engineers, and construction providers—both contractors and construction managers—to ensure that the primary purpose is met. Whether this is a new facility replacing on-site treatment or upgrades to an existing facility, there are often limited funds available for construction and for the operations and maintenance of these facilities. Careful attention to site conditions, equipment specifications, project delivery method, constructability, ease of operations and maintenance, and training in the planning and design stages will allow these limited funds to be used most efficiently and effectively.

Furthermore, it is important for everyone involved in the project as well as city officials to understand what should happen during the construction and startup phases, and the type of staff and services that are needed. This manual will help guide these officials and the facility staff and regulatory agencies through these phases of a project, and give information on how to manage the complexities involved with these project phases.

In this manual, a successful project is defined as one in which the upgraded facility meets all effluent and state regulatory agency certification requirements; the equipment is easy to operate and maintain, and is reliable; construction is completed on schedule and on budget; and the owner, engineer, and contractor are talking to each other at the end of the project.

The authors and reviewers of this Manual of Practice represent a cross section of the industry, including consultants, operators, owners, and contractors, and have been through many upgrade projects from start to finish. They have undertaken multiple projects for a range of clients and represent many years of combined experience in the field.

2.0   USE AND ORGANIZATION OF MANUAL

This manual is useful for projects at both large and small treatment facilities. At large or complex treatment facilities, everything discussed in this manual may be required. For small facilities, only portions of this document may apply. This is also true for a major upgrade compared to a unit process upgrade. Furthermore, although it is geared toward municipal water resource recovery facilities (WRRFs), everything in the manual is applicable to water or reuse treatment facility and industrial treatment facility upgrades. Although the specific design approach, phase name, and number of design steps may vary from consultant to consultant and differ from those mentioned in this manual, the final product should have the basic sequence and elements discussed here.

This manual is organized in the same way as a project would be organized, beginning with Chapter 2, "Planning Upgrades to Existing Facilities," which describes what happens during the early planning stages. Chapter 2 provides a high-level overview of the need for and reasoning behind planning for projects to implement a new WRRF or for enlarging and upgrading an existing facility. Examples of considerations for parties during this stage include:

- What are the drivers for the upgrade?
- Is the upgrade a regulatory requirement because of the age of the treatment facility or new water quality requirements, the economic value of producing reclaimed water for beneficial use, increased growth and development in the community, or a combination of these factors?
- What is the purpose of the plan? Who is the audience, how is it developed, and what should it contain?
- How does one procure the design engineer?
- What qualifications should one be looking for during the procurement process?
- What should be considered in drafting contract terms?

Chapter 3 is a new chapter, not included in the first edition of this manual, to address a specific situation: replacing rural, dispersed treatment systems such as septic tanks with regionalized collection and treatment in more complex systems. Generally, this situation is identified in a regional water quality or master plan, which serves as the foundation for planning, design, and implementation of these larger systems.

Chapter 4 describes the facility plan in detail. Chapter 4 specifically addresses the components and procedures of a facility plan. The facility plan is the primary document that describes the planning and decision-making process that leads to improvements or new construction of WRRFs. It also includes the implementation schedule, project delivery strategy, major milestones, potential off-ramps, and cost estimates. The facility plan is typically the first phase in the upgrade process; however, not all projects require a full facility plan. Other engineering reports or smaller facility plan amendments are often the only requirement for a facility upgrade project. This is probably the most important phase because it sets the stage for the design process and helps to determine the various funding sources that will be available. The facility plan is developed to establish the goals of a project and outline the general processes required to achieve that goal. The facility plan should be linked to and flow from the strategic and master plans of the organization and should be supported by a long-term financial plan. The strategic and master plans should be consulted for policy guidance, design criteria, and standards.

Chapter 5 is an overview of various project delivery methods. In the past, most utilities hired an engineer for planning and design; the project was then bid, and a contractor was selected. The design engineering firm would provide project oversight and resident engineering and inspection during construction. Now there are alternative project delivery methods, which may make these projects more efficient and economical; however, sometimes these can make the project less efficient and costly. In many localities, state law controls the use of these alternative project delivery methods for public projects. Even in states that allow these, some communities may have procurement policies that prohibit alternative delivery methods. If that is the case, and these alternatives may be attractive to your project, you might have to take steps to change procurement policies. This can take time, so it is important to have as much knowledge about them as possible. This chapter presents an overview of the many project delivery methods now available to public and private entities.

Because there are so many project delivery methods available, it is important to understand how the program will be structured so that the subsequent phases of design and value engineering can be planned and scheduled. The traditional "bid phase" of a conventional design–bid–build delivery method was addressed in a separate chapter in the previous edition, but is now included in the discussion on project delivery methods.

Chapter 6, "Value Engineering and Constructability Reviews for Facilities," is now presented before the design steps because value engineering should start, or be considered, at the beginning of a project. The project may be small enough so that value engineering is not required, but it is always good to keep the concept of providing value at the forefront because value engineering can be applied at various levels of both planning and design. This chapter describes value engineering, what it is, at what stages in the design process it should be conducted to be most beneficial, what qualities one should look for in the value engineer, and what should be included in the scope of services. In many states, the regulatory or funding agency (or agencies) involved have specific value engineering requirements. These dictate at what stage and how many times value engineering should be conducted. Chapter 6 provides the information needed to ensure that regulations are followed.

The next set of chapters concentrate on design. Chapter 7 describes the preliminary design phase. This is the phase in which the concepts described in the facility plan are evaluated and analyzed further. This is an extremely important phase in which mass balances and process models are developed, maintenance of facility operation during construction is evaluated, and the permits required before final approval and construction are identified. The chapter discusses defining the project; developing the plan for the contract documents (drawings and specifications); and finalizing the concepts developed during the facility plan, including design criteria, process selection, and safety and security requirements. This chapter also includes discussions on determining various constraints and requirements, defining various standards and styles, and, finally, managing the effort.

Chapter 8 details the final design phase. It describes how this phase is managed, and the participants and their responsibilities; provides recommendations on reviewing the contract documents, and other agencies that should be involved with the review; and discusses the effect of construction activities on the business and residential community. In addition, the chapter discusses finalizing and confirming the following items: process selection, including process flow diagrams; hydraulic profile; selection of major equipment; site layout; and developing the cost estimate. Finally, there is a discussion on preparing construction documents, including preparation of lists and schedules, providing specifications, and developing a construction schedule and operating plan.

Chapter 9 describes the construction phase of the project. This chapter discusses the scope for the engineering services for this phase, what services should be included, on-site engineering supervision requirements, change order management, facility shutdowns, and other issues that arise during construction. The chapter also includes sections about what happens after construction starts; facility manager and staff roles; how often the parties should meet and who should attend job meetings; and when training should occur, who should be trained, and how much training is required. There is also a discussion of regulatory requirements at the end of the project, how best to deal with punch lists, and when and under what conditions final acceptance of the facility and equipment should occur.

Chapter 10, "Facility Startup and Commissioning," discusses some key issues involved with the completion phase and the start of ongoing operations. The chapter discusses who should be involved, what happens if things do not work (either process or equipment), and warranties for equipment and the advantages of extended warranty through maintenance bonds. Also included in this chapter are comments concerning the startup phase. Startup really begins with equipment testing and checkout by the contractor before being taken over by facility staff. This is a function of a well-thought-out startup plan prepared by the facility manager in cooperation with the contractor, construction manager, resident engineer, resident project representative(s), and facility staff. This plan will define responsibilities and responsible parties, identify equipment and tanks that will be placed into service, and explain starting sequence, emergency response, and troubleshooting, to name a few.

Whether you are a utility manager, operator, consultant, or contractor, use this manual wisely. It is a guidance document, not a hard-and-fast set of rules and regulations. Many situations are specific to the site, community, state, and/or region, so a hard set of rules cannot be developed. Use the manual to think about your individual project to make it the best project it can be. The authors of this manual have tried to touch on every subject that should be evaluated during the upgrade process. The goal of the authors is to ensure that projects are run efficiently and economically and that we can all learn from each other and not make the same mistakes.

2

Planning Upgrades to Existing Facilities

Yanjin Liu, PhD, PE; Jianfeng Wen, PhD, PE; & James Chelius, PE

1.0   IDENTIFYING THE NEED FOR UPGRADING A WATER RESOURCE RECOVERY FACILITY

A facility plan or comprehensive planning study presents a strategy for facility improvements to ensure that utilities can continue to provide safe, adequate, and reliable service to the community and public. A scope of work should be clearly defined before conducting a facility plan. In a scope of work document, the need for upgrading a water resource recovery facility (WRRF) will be identified, and the work that will be performed needs to be summarized. Typically, a facility plan will conduct a detailed evaluation and assessment for present and future conditions of the WRRF. A minimum 15- to 20-year planning period is a reasonable timeframe. The needs identified in the scope of work should be discussed thoroughly. Alternative solutions will be evaluated, and the best option should be recommended in a facility plan. The necessary improvements will serve as the basis for the detailed design and construction of each capital improvement project.

2.0   DRIVERS FOR UPGRADING A WATER RESOURCE RECOVERY FACILITY

The reasons for conducting a facility plan depend on many factors. The inability to meet a stringent effluent discharge limit will drive a WRRF to an upgrade of treatment processes to meet regulatory requirements. A facility plan can be conducted to increase treatment capacity and upgrade equipment needed to address population growth and aging infrastructure. It is more common for a WRRF to address multiple needs in a facility plan. Some common drivers for upgrading a WRRF are identified and discussed in this section.

2.1   Regulatory Requirements

A WRRF shall be in compliance and in conformance with the discharge permit—and even with reclamation criteria—to meet regulatory requirements. Nationally, there has been a strong trend toward more and tougher regulations affecting facility effluent quality. Examples include more stringent water quality regulations, additional regulation of treatment facility residual management, and guidelines on reclaimed water reuse. Typically, these new regulations are passed down from the federal to the state level, where they are incorporated as state laws. Often, states and regional river basin commissions have passed additional legislation to address issues specific to their individual circumstances, and this can be more stringent than federal laws.

Therefore, in developing a facility plan, it is beneficial to review national, state, and local regulations that can affect future planning. In this section, the regulatory requirements will be discussed in terms of the following three perspectives: water quality requirements, violation of an existing permit, and consent decrees. They are the three primary situations when conducting a facility plan may be necessary to meet regulatory requirements.

2.1.1. Water Quality Requirements and Changes to Water Quality

Stringent water quality requirements are often the main driver of facility upgrades. The goal of a WRRF is to continue to collect and treat wastewater to levels that meet or surpass federal and state water quality standards. The discharge permits for each WRRF are regulated by the National Pollutant Discharge Elimination System (NPDES) permit. Treatment facilities are designed to meet NPDES permit requirements under projected flows and loads, and to comply with water quality regulations. If new water quality parameters (e.g., total phosphorus) and/or stricter effluent discharge limits (e.g., total nitrogen) are expected to be implemented in the near future, a facility plan or engineering study should be conducted to evaluate the effects of new and stricter permits on treatment processes. In the facility plan, the treatment performance of a WRRF shall be evaluated by comparing the treated effluent quality with existing and future permit limitations. Recommendations for capital improvements should be offered to the WRRF to meet the new discharge limits if existing treatment processes are not capable of doing so after evaluation. This will ensure continued compliance with existing and anticipated federal and state water quality and environmental regulations for the WRRF.

2.1.2 Violation of an Existing Permit

Violation of an existing permit can result in the issuance of a Notice of Violation (NOV) to the facility. A facility plan and upgrade may be needed to address NOVs and meet the NPDES effluent discharge limits in the future. Violation of an existing permit may be the result of insufficient treatment capacity, and thus an expansion of the treatment facility may be needed. The violation may also result from an inability to meet water quality parameters under either monthly average or daily maximum conditions. The existing treatment process needs to be upgraded to comply with the regulations and discharge permits. Upon completion of the upgrades, the facility is expected to meet its NPDES permit requirements under monthly average and daily maximum conditions. However, sometimes NOVs can also be the result of operational issues, mechanical equipment failure, and other causes that may not result in large capital needs. Therefore, an investigation should be conducted to identify the root cause of the NOV and determine if a facility plan is needed. Violation of an existing permit is typically a significant driver for requiring a facility plan or engineering study. Capital improvement recommendations should be proposed if needed to address these issues and ensure that the WRRF complies with its discharge permit in the future.

2.1.3 Consent Decrees or Orders

A consent decree or order is a legal document that formalizes a settlement agreement between the regulating agency and the permittee. The consent decree is designed for the most cost-effective use of resources to improve the WRRF and collection system within a reasonable timeline. The regulating agency and the permittee will negotiate the elements of the consent decree to reach an agreement on resolving environmental violations. A facility plan or engineering study is typically included in a consent decree. Certain tasks must be completed within a specific timeframe to fulfill the decree requirements. For example, a consent decree may require the permittee to conduct a facility plan to cover a 15-year planning horizon. This plan will assess and recommend improvements to aging wastewater lines and infrastructure so that sanitary sewer overflows may be reduced. A penalty is typically associated with the decree document if the permittee violates the consent decree.

2.2   Growth and Development in the Community

Growth and development in the community accounts for a significant reason to expand and upgrade a WRRF. Comprehensive facility plans should provide population and flow projections for a minimum 15- to 20-year planning horizon to evaluate the need for facility expansion because of increased flows. For service areas experiencing rapid population growth and increasing population density, there may be a need for facility expansion to increase treatment capacity to treat additional flows that will be generated as a result of increased development and density in the near future. For other service areas, the growth and development may be relatively slow, and the need for facility expansion may be an issue for the long-term. A commonly used rule for determining the need to expand a facility is that the current and projected average daily flow should be less than 80% of the rated (permitted) capacity of the treatment facility. If average daily flow is currently or projected to be greater than a certain threshold (e.g., 80% or 90% for some states) of rated (permitted) facility flow, a detailed analysis of the capacity of each of the processes at the facility should be performed in accordance with the 10 States Standards and/or other state or local guidelines to help plan for potential expansion, with any recommended improvements to meet projected flows implemented within the planning horizon. This will help the WRRF keep pace with population growth, lower the risk of facility performance issues and potential discharge violations, and ensure reliable wastewater service to the community into the foreseeable future. The effect of climate change on population and community growth should also be evaluated because increasing levels of infiltration and inflow may affect influent flows (please see Section 2.5 on asset resilience).

2.3   Regionalization and Decentralization

Regionalization and decentralization are other drivers for needing a facility plan. Regionalization can often provide economics of scale, avoid duplication of facilities, and prove to be more effective service to customers. For example, facilities within a specific geographic area can be consolidated to benefit from the shared collection system and treatment facilities. Consolidation of wastewater flows with one new lift station will improve reliability and reduce operations and maintenance (O&M) costs. The regionalized facilities may significantly reduce labor costs, O&M for equipment, and save overall energy and chemical usage. Regionalization can also enhance the potential for resource recovery (digester gas, power generation, etc.), which may not be practical or economical at smaller facilities. Alternatively, decentralized wastewater treatment can sometimes meet the triple bottom line goals more effectively to protect the environment, reduce financial effects, and minimize social burdens. For example, a package wastewater treatment facility that is used to treat and dispose of relatively small volumes of wastewater may be most efficient for a small community. A detailed discussion on decentralized facilities is available in Chapter 3 of this manual. Overall, regionalization or decentralization opportunities should be evaluated in the facility plan to ensure continued protection to the environment and human health in a cost-effective manner.

2.4   Aging Infrastructure

Identification of old equipment and aging infrastructure is one of the most common reasons to conduct a facility plan and trigger a facility upgrade. Typically, aging infrastructure in poor condition and near the end of its useful life is unable to provide reliable performance, lacks sufficient capacity, and often requires intensive maintenance to be kept in working condition. Failure and breakdown of the equipment may occur and interrupt operations of the WRRF. This poses a risk for the facility in meeting water quality regulations. Equipment with outdated technology reduce O&M efficiencies. For example, old blowers that are not equipped with variable-frequency drives may result in increased energy consumption for the facility. Outdated ultraviolet equipment also consumes more energy than current technology while achieving the same performance level. A facility plan provides a detailed review and evaluation of the condition and performance of aging infrastructure. It also recommends solutions and suggests capital improvement projects to upgrade deteriorating assets, increase reliability, increase hydraulic capacity, and reduce the risk of permit violations.

2.5   Asset Resilience

In recent years, there has been increasing interest in incorporating resilience into the wastewater management of public utilities, which can be another driver for conducting a facility plan. In the northeastern United States, Hurricane Sandy in 2012 demonstrated how extreme storms can greatly affect wastewater treatment facilities. Sea level rise is particularly problematic, as facility outfalls often discharge into tidally influenced water bodies and facility infrastructure can be vulnerable to storm surges exacerbated by sea level rise. Options such as relocating facilities, building flood walls, elevating and providing adequate redundancy for critical assets, and increasing facility capacity where appropriate can greatly reduce WRRF vulnerability to extreme weather events and keep system disruption to a minimum. WRRFs need to be assessed against the new norm of more intense, more frequent storm events to identify vulnerability and timely mitigation measures that will improve asset resilience and ensure continuous service under such conditions.

2.6   Moving to Resource Recovery

Wastewater is increasingly considered as a resource rather than a waste product. Diverse resources (e.g., reclaimed water, nutrients, energy) can be recovered from wastewater. The trend is to upgrade existing conventional wastewater treatment facilities to WRRFs to achieve cost savings and meet more stringent discharge limits in a sustainable manner. Three primary elements that drive the need to upgrade to WRRFs are energy conservation and recovery, resource recovery from the liquid stream, and resource recovery from the solids stream. These are briefly discussed below.

2.6.1   Energy Conservation, Recovery, and Production

Wastewater treatment is highly energy intensive because energy is used throughout the treatment processes. Aeration and pumping are typically the largest energy users at a WRRF. Energy costs are also rising as a result of aging infrastructure, more stringent effluent requirements, and increasing electricity rates. Wastewater utilities should develop and implement energy management strategies to conserve and recover energy. A comprehensive energy audit at the facility can help to establish current energy consumption patterns and identify opportunities for energy savings. An energy audit examines energy usage of each part of the treatment process and determines the most energy-intensive processes and equipment. The baseline assessment information can then be used to evaluate alternatives for process and equipment changes to improve overall energy efficiency. Replacement of energy- and maintenance-intensive blowers to energy-efficient blowers reduces energy consumption and improves O&M efficiency.

WRRFs have been striving to evaluate and identify opportunities to use renewable energy sources such as biogas, solar, wind, hydropower, or thermal energy to fully realize the benefits of wastewater as a resource. Among those opportunities, biogas produced in the anaerobic digestion process is recovered at many WRRFs as a renewable and sustainable energy source for energy production. It is estimated that approximately 40% of the energy needed for aeration in a typical activated sludge process can be offset by the energy recovered from combined heat and power production. WRRFs are transitioning from conventional energy-intensive wastewater treatment processes to zero-discharge, energy-neutral, or even energy-positive wastewater treatment processes.

2.6.2 Resource Recovery from Liquid Stream Treatment

Treated wastewater for reuse and recycling provides a combination of water supply and wastewater management benefits by collecting nonpotable water, including treated or raw wastewater and stormwater, and then treating it as needed and distributing it for beneficial reuse. The growth in recycled water use results from the growing demand for water and water supply shortages, stringent regulations on wastewater discharge limits, the increasing cost of water treatment upgrades, development in remote areas, enhancement of treatment technologies, and the desire for sustainable development. It is important to identify and incorporate opportunities for expanding water reuse applications into the planning for WRRFs to implement effective water resource management.

2.6.3 Resource Recovery from Solids Stream Treatment

Biosolids produced at wastewater treatment facilities are nutrient-rich organic materials. Technologies have been developed and implemented at a number of WRRFs to recover nutrients (phosphorus and nitrogen) from solids stream treatment. Life cycle analysis can be used to identify economically attractive alternatives by taking into account a wide range of factors such as capital investment, O&M costs, regulatory conditions, and social and environmental values.

2.7   Other Drivers

Besides the drivers discussed previously, there are additional drivers (e.g., safety, security, O&M efficiency) for conducting a facility plan. For example, if a facility building is in poor condition, it would put the facility at risk of not being able to provide a safe working environment for operators and would also affect wastewater treatment reliability. A facility upgrade may be needed to repair or rehabilitate the building to improve safety. Other safety-related upgrades for a facility include installation of grate covers and handrails for treatment tanks and wet wells. These upgrades protect operators working in the WRRF. Security upgrades should also be considered to protect critical assets by installing fencing and/or electrical gates to improve security and reliability. Higher levels of cybersecurity should also be considered for integration into treatment process control. An increase in O&M efficiency in chemical use or automation and control could also drive a facility upgrade. The drivers to upgrade a facility depend on many factors and should be evaluated based on facility-specific conditions and performance goals.

3.0   WHAT IS THE PURPOSE OF FACILITY PLANNING?

The primary purpose of a facility plan is to provide an engineering analysis and a comprehensive evaluation of the existing WRRF's treatment infrastructure to determine short-term and long-term improvements required for a utility to continue to provide high-quality service to the community and public in compliance with all regulatory standards. The development of a short-term strategy includes a review of known or imminent problems in the system, capacity analysis, and a review of development currently approved or planned over the next several years.

Typically, the short-term plan addresses the following components:

- Needed improvements to maintain compliance with existing regulations
- Infrastructure improvements required to provide sufficient capacity to convey sewer flows and support new developments on the current planning horizon

- WRRF upgrades to address any capacity limitations and/or condition issues

The long-term plan addresses anticipated system capacity issues resulting from expected increases in total sewer flows from the following:

- Build-out within the service area
- Asset resilience improvement
- Regionalization and/or decentralization of treatment systems

A comprehensive facility plan details the capital improvement recommendations for WRRFs for the next 15 to 20 years. The plan presents a strategy for facility improvements to ensure that WRRFs continue to provide safe, adequate, and reliable service to the public. The contents of a facility plan report could include the following:

- Executive Summary: provides a summary of the key findings and includes a prioritized list of recommended projects
- Compressive Planning Process: provides an overview of the comprehensive planning process used in the facility plan
- Flow and Load Projections: provides population, flow, and load projections for the next 15 years for a WRRF.
- Facility Plan: addresses the following for each system:
  - System overview and general information
  - Detailed description of treatment processes
  - An assessment of treatment performance and compliance with current and future NPDES discharge requirements
  - Capacity analysis and comparison of flow projections to facility capacity
  - Condition assessment of major treatment processes, equipment, and buildings
- Recommendations: provides more detailed descriptions of each recommended project for a capital improvement plan, including the need for a project, project background, recommended solutions, benefits of the project, alternative analysis, and budget discussion
- Appendix: provides additional information to support the facility plan, including a cost estimates breakdown for each recommended improvement project, flow monitoring data, and water quality data

4.0   HOW IS A FACILITY PLAN DEVELOPED?

A facility plan for a WWRF can be developed with two different approaches: procurement of an engineer or consultant or using the owner's in-house resources (this will be briefly discussed in Section 5 below). If the facility plan is developed by an engineer or consultant, then it is developed through a scope of services specified by the owner (municipality or wastewater utility). During this process, the deliverables of the facility plan should be developed with input from the owner, thoroughly reviewed, revised as needed, and finally approved and adopted by the owner. The details of how to procure an engineer are discussed in Section 6 of this chapter. The detailed procedure on how to develop a facility plan will be discussed in Chapter 4.

5.0   OWNER'S IN-HOUSE APPROACH

Depending on the owner's needs, objectives, scope of work, timeline requirements, and availability of resources, some utilities and municipalities may choose to develop a facility plan using an in-house approach. The benefits of conducting facility planning internally can include, but are not limited to, better understanding of facility needs and requirements, retaining knowledge within the organization, better management and control of the process, and cost efficiency. Similar to the approach of using an outside engineer, the facility plan can be developed by relying on internal resources and following these steps:

- Prepare the scope of work and budget
- Allocate resources, including staff and budget
- Conduct facility planning
- Prepare recommendations
- Deliver the facility plan and conduct a final review meeting with internal and external stakeholders

6.0   HOW TO PROCURE AN ENGINEERING FIRM

The majority of WRRF owners hire an external engineer or consultant to conduct wastewater treatment facility planning and design. The process of procuring an engineer varies from state to state, region to region, and utility to utility. However, a guidance principle for the hiring process, termed "qualification-based selection" (QBS), has been developed by federal government agencies and adopted by many state governments (American Institute of Architects of Illinois, American Council of Engineering Companies of Illinois, Illinois Professional Land Surveyors Association, & Illinois Society of Professional Engineers, 2000; American Council of Engineering Companies of Massachusetts, 2006). It is important for owners to follow the QBS principle to procure engineers who are best qualified to perform the task and prepare a high-quality facility plan. Chapter 4 of this manual will discuss the details of how to develop a facility plan from a technical perspective.

6.1   The Owner's Role in the Process

If the facility plan is conducted by an engineer or consultant, the owner's facility manager and operation staff should be actively involved in all phases of the project. The engineering firms rely on the owner to provide input and information regarding facility O&M. The owner's views and preferences will guide the engineer on the treatment technology selection process; the owner will also be responsible for administering the project and overseeing the consultant's work. The owner will review the facility plan and other documents prepared by the consultant and provide comments and feedback. Some owners may choose to take on some of the tasks described in Chapter 4 during the facility planning, depending on staff expertise and availability. It is important that the role of the owner is clearly defined at the beginning of the process so that every party is in agreement to avoid unnecessary confusion during the process.

6.2   Qualifications-Based Selection Criteria

As codified in the Brooks Act (Public Law 92-582) in 1972 to protect the public's health and safety and protect the taxpayers' interest, QBS is a competitive procurement process in which the owners select the engineering firms using their qualifications as the criteria, not solely cost. QBS has been successful not only at the federal level, but has also been adopted by 44 states and hundreds of local municipalities throughout the country. The QBS process is widely endorsed by the American Council of Engineering Companie, the American Public Works Association, the Associated General Contractors, the American Bar Association, and all major design professional associations. The QBS process typically includes the following steps:

1. The owner identifies the general scope of work and project timeline.
2. A request for qualification is issued and a statement of qualifications is evaluated from interested firms.

3. A short list of qualified firms to be interviewed is determined.

4. Interviews are conducted and the firms are ranked.

5. Negotiations are conducted with the top-ranked firm. If an agreement cannot be reached with this firm, the negotiations are terminated and the owner enters into negotiations with the second-ranked firm and so on, until an agreement is reached.

The step-by-step procedure is discussed in detail below.

6.2.1   Requests for Qualifications

As required by the Brooks Act, the QBS process typically begins with the owner's issuance of a request for qualification (RFQ) and announcement of interest to receive proposals from engineering firms for a particular project. This RFQ should include, but is not limited to, project overview, objectives and requirements, procurement process, statement of qualifications, evaluation process and criteria, and submittal requirements. Because engineers are selected through this qualification-based process, the owners typically ask the engineers to provide the following information:

- Cover letter
- Brief description of the firm
- Project understanding and approach
- Experience statement identifying similar projects and areas of expertise
- Project team chart (with resumes) for personnel involved in the project

6.2.2   Interview Process

Upon receipt of the proposals and reviewing the firm's qualifications, owners develop a short list of three to five firms for interviews. A tour of the facility is sometimes hosted by the owner before the interview to give the firms an opportunity to better understand the facility and ask questions. Such facility tours also allow owners to gain initial feedback from the engineers. Then the owners will invite the short-listed firms to give a presentation to the owner's selection committee. Before the interviews, a ranking tool should be provided to selection committee members and a preliminary ranking should be completed. The selection committee typically includes staff with various expertise related to the project. Common criteria used in the selection include:

- Firm's understanding of the project requirements and needs
- Qualifications of the firm or individuals, including project manager and key personnel

- Experience on similar projects
- Project approach and understanding of the project
- Use of subconsultants or contractors
- Schedule control
- Financial stability of the firm
- Depth, location, and availability of staff
- Communications approach

The owner should keep in mind that the individuals interviewed may not be the same ones who will do the majority of the work, and it is important to understand the roles and level of involvement of each individual on the firm's team. After all the firms have been interviewed, the selection committee will meet and discuss their impressions and rating of each firm and agree on a ranking of the most qualified firms. The top-ranked firm will be moved forward to begin a negotiation process to award a contract. The second- or third-ranked firms can be used as alternates if negotiation is not successful with the top-ranked firm.

A sample "Interview Questions and Score Sheet" is included Table 2.1. The selection committee should mutually agree on the rating for each criterion before completing the form.

6.3   Financial Considerations

The owner will begin to negotiate with the top-ranked firm regarding the scope of work and an engineering fee that is reasonable and justifiable. The owner is also responsible for providing information on any activities that might affect the schedule and the consultant's work. The owner should request the following fee proposal with the agreed-on scope of work from the selected firm:

- A breakdown of costs for each associated labor category, labor hours, and labor costs for each phase of the project
- Cost of any proposed subconsultants
- All other direct costs (travel, equipment rental, laboratory analysis, etc.)
- The engineer's overhead rate
- Proposed profit
- Firm's insurance

If an agreement cannot be reached with the first-ranked firm, the owner terminates the negotiation and will begin negotiating with the second-ranked firm and so on down the list until an agreement is reached and a firm is selected.

TABLE 2.1   Interview questions and score sheet (adapted from the American Institute of Architects of Louisiana, 2018)

Interview Question and Score Sheet				
Owner				
Project				
Categories		**Rating (1–5)**	**Weight (1–10)**	**Total**
1	Related project experience			
2	Firm's ability and capacity to perform the work			
3	Grasp of project requirements			
4	Method to be used to fulfill the required services, including design and construction phases			
5	Management approach to technical requirements			
6	Use of consultants that may work on project			
7	Time schedule planned for project			
8	Firm's experience and methods used for budgeting and financial controls determining fee and compensation			
		Grand Total =		

Instruction for the Interviewers:

Rating: During the interview, rate each firm on a scale of 1 to 5 in each of the eight categories.

Weight: Weights on a scale of 1 to 10 should be established for each category before the interview. It is suggested that weights used here correspond to weights of categories used for evaluating the Letters of Qualifications. Enter the pre-established weight for each category on the lines provided.

Totals: At the completion of the interviews, multiply the rating by the weight in each category and enter the total on the lines provided. Add all totals to establish the grand total.

Interviewer:

Firm:

6.4   Scope of Work

The scope of work varies depending on the extent of the specific requirements of the project and the phases of the project (i.e., study, design, construction, field services). The engineer's proposal typically includes both the "basic" and "additional" services required to complete the project. Basic services are those directly related to core engineering disciplines and for which the scope of work and level of effort can be well defined. Basic services typically include structural, process, and electrical engineering, as well as architecture, HVAC (heating, ventilation, and air conditioning), and plumbing. Additional services generally include surveys, subsurface investigations, geotechnical evaluations, permitting, administration, public participation programs, and construction observation.

The selected firm should be asked to prepare a detailed fee estimate predicated on the agreed-upon, detailed scope of work as the basis of compensation negotiations. As an integral part of the scope discussions, the requirements and format of the proposed written contract must be discussed. The owner and design professional may wish to use the standard forms of agreement used by professional societies.

6.5   Form of Establishing a Contract

The agreement between the owner and firm must ensure that both parties have the same expectations and understanding of the project requirements. Legal counsel should be involved in final formulation of the agreement for securing engineering services. Below are several commonly used types of contracts:

- Lump sum: A lump-sum contract is one of the most common engineering contracts, and is used when the work is clearly defined to allow the engineer to provide an accurate assessment of the level of effort. The engineer is hired for a fixed sum, and the cost of the services will not exceed the contract amount unless the owner requests the engineer to provide services that are beyond the original scope of work. This type of contract often pertains to projects that are simple and have little or no chance of evolving while the project is underway.

- Cost-plus fixed fee: The cost-plus contract is used when the owner agrees to pay for the labor, materials, and an additional amount for the engineer's indirect costs, overhead, and some profit. This payment method is commonly used to pay for the engineer's field staff during the construction phase of the project.

- Unit price: The unit price contract is based on estimated quantities of items included in the project and the unit price of each item. The final price of the project often depends on the quantities of each item required for the project. This type of contract is only used for construction and supplier projects where the different types of items can be accurately identified. Because of this restriction, it is not uncommon to combine this type of contract with the lump sum contract or other types of contracts for engineers.

After the contract is finalized and the owner and the engineer have signed the contract, the selection process is complete and the project begins.

7.0   REFERENCES

American Council of Engineering Companies (ACEC) of Massachusetts. (2006). *Qualifications based selection owner's manual.* https://files .engineers.org/file/QBSMAN-10-2006.pdf

American Institute of Architects (AIA) of Louisiana. (2018). *Qualifications-based selection workbook for state and local governments.* https://aiala .com/wp-content/uploads/QBS-Workbook.pdf

American Institute of Architects (AIA) of Illinois, American Council of Engineering Companies (ACEC) of Illinois, Illinois Professional Land Surveyors Association, and Illinois Society of Professional Engineers. (2000). *Qualifications-based selection: A guide including model local government policy and procedures for selecting architects, engineers and land surveyors.* http://docs.acec.org/pub/9E675727-0EEE-1DC9-3B51-2A94F3CFDF3B

3

Upgrading from Decentralized Facilities

José Velazquez, PE, BCEE

1.0 INTRODUCTION

Historically, the handling of wastes produced from human activities has progressed from smaller, decentralized systems, ranging from pit privies to outhouses, to on-site wastewater treatment systems (OWTS), cluster-type community systems, and smaller neighborhood or limited-service areas. Community handling systems may ultimately become regional mechanical treatment facilities to meet the demand in small geographic areas with high population densities. It is also possible that smaller neighborhood systems are eventually abandoned or regionalized within a central wastewater treatment system. For a more complete summary of how collections systems and centralized treatment have progressed over time, see the History of Sanitary Sewers website (www.sewerhistory .org). A good description of an OWTS is provided on the U.S. Environmental Protection Agency (U.S. EPA) website under "Types of Septic Systems" (U.S. EPA, 2021). Drivers for this progression include densification of development, land-use changes, and protection of groundwater quality—especially in areas that rely on groundwater wells to provide potable water with little or no treatment.

The concepts, processes, procedures, and systems described in this manual of practice apply to construction of both new, "greenfield" water resource recovery facilities (WRRFs), and to existing facilities that are being upgraded or enlarged.

2.0 CONSIDERATIONS FOR REGIONALIZATION

As areas grow and populations increase, there is often a need to move from individual, less-involved wastewater treatment/water reclamation facilities to more complex, centralized facilities. Water quality considerations may also drive the change to systems that provide higher levels of treatment while directing treated water to other end uses.

2.1 Who and What Drives the Decision

The decision to move from an OWTS to a centralized system is complex and includes many factors beyond finding and implementing a potentially viable technology for the community. Drivers that may create the need to consider regionalization include:

- Private owners' preference
- Jurisdictional (town, city, county, etc.) requirements, including sewer use ordinances and results of annexations
- Owner or developer's desire to increase housing density
- Regional water quality and source water protection
- Regional water reuse goals
- Operations and maintenance (O&M) costs
- Physical constraints, such as topography and geology
- Need for multiple facilities, such as wastewater lift stations or smaller, more dispersed treatment facilities
- Overall life-cycle costs for collecting transmitting, treating, and reusing the water

Because most individual systems are owned by a single property owner or a few owners, one of the first decisions is to adopt a management model that provides services to multiple users through a centralized management structure. This can take the form of an owner's association, an improvement district (such as a special improvement or sanitary improvement district), existing government-owned and operated treatment provider (e.g., city or town public works or wastewater services department), other regional providers, or privately held entities. When regionalization is considered, another decision to be made centers on how the utility will be funded, along with how it will be managed, operated, and maintained. This requires adequate resources to ensure that the services provided are sustainable and meet the needs of the community, including financial, management systems, and O&M staff.

New regional facilities not only provide environmental protection, improved water quality, and opportunities for producing reusable resources (water, biosolids, and others), they can also become a community resource for educating the public and demonstrating environmental stewardship.

2.2   Applying a Step-Wise Decision-Making Process

Implementing a new WRRF/system will follow a path typically defined by the regulatory entity in which the facility is located. The details of each step will be defined at the local level. However, it is important to understand the steps that will be needed. Typically, these include developing legal authority, establishing funding systems, involving stakeholders, developing capacity, and then implementation.

3.0   IDENTIFYING AND SECURING A SITE

Considerations must be made regarding location in the service area, space needed, land uses, buffer zone requirements, access to the site and other utilities, aesthetic considerations (i.e., odors, visual, noise), security, and environmental considerations (land, air, and water resources).

Regardless of the final uses for the various resources, such as treated effluent and biosolids, regulations for protecting air, land, and water must be followed. The project owners must comply with applicable federal, state, and local regulations. All facilities, whether new or an upgrade, must comply with effluent discharge regulations.

If a site is to be acquired, then land acquisition processes will be followed. These may include changes in planning and zoning, and public notice and stakeholder communications. Site constraints such as access, subsurface conditions, floodplain impacts, seismicity, adjacent land uses, historical context, and public input from adjacent and neighboring properties will need to be addressed. To acquire property, a conceptual level of design should be available to identify the area needed and the general aesthetics of the facility, and for use in zoning changes. The availability and cost of land can be a significant factor in the overall process and must be addressed in the early stages of planning.

4.0   WATER RESOURCE RECOVERY CONSIDERATIONS

The most common application of an OWTS is in homes whose potable water supply is from an individual well serving one to a few homes. The wastes produced from homes in systems of this size can be managed using individual or community systems, or by regional providers. The homes that rely on OWTS-based waste management are, in fact, providing localized reuse by returning the water used in the home to the groundwater source. As more users are served, other larger and more regional reuse systems can be considered.

5.0   SERVICE AREA UPGRADES

Elements of the project that should be considered to ensure water reclamation services are provided equitably throughout the service area include gravity collection systems piping, interceptors, pumping stations, force mains, and ancillary equipment such as odor control, telemetry, instrumentation and control (I&C), and supervisory control and data acquisition services.

Physical improvements that may be needed in the service area include wastewater collection piping (both gravity and pressure lines), pumping facilities, odor control, flow monitoring, and sampling equipment. In areas that have existing development with OWTS, wastewater collection may be provided by installing septic tank effluent pumps (STEP); existing septic tanks remain in service and the effluent from them is collected and pumped, eventually to the treatment facility. Installing piping systems requires obtaining land or right-of-way for the facilities. This process will involve property acquisition, surveying, and legal work. It is also important to obtain adequate area to construct and operate the systems. Finally, consideration must be given to providing supporting infrastructure. Auxiliary utilities that may be needed include power, I&C, access, site security, and stormwater management and erosion control facilities.

At the same time the physical assets are being discussed, studied, and defined, the asset-owning entity may need to be developed. The administrative capacity of the entity owning the water reclamation system must be able to handle the expected number of customers, communicate with those customers, and hire and retain multidisciplinary staff members. These staff will be required to operate the facility, provide maintenance for mechanical equipment and piping systems, and support the utility in managing its assets (including collections and treatment systems) while generating adequate revenues to pay for services, cover debt, and budget for future needs. Consideration must be made regarding the size and complexity of these systems so that the support provided meets the needs of the customers, operators, administrators, and stakeholders without excessive complexity or financial cost. In short, the support structure should be "right-sized" for the utility.

6.0 REFERENCES

U.S. Environmental Protection Agency. (2021). Types of septic systems. https://www.epa.gov/septic/types-septic-systems

4

The Facility Plan

Jose Christiano Machado Jr., PhD, PE; Keli Callahan, PE;
Nicole Stephens, PE; & Jenny Hartfelder, PE

1.0   INTRODUCTION

The facility plan is the primary document that describes the planning and decision-making process that leads to improvements or new construction of water resource recovery facilities (WRRFs). It also includes the implementation schedule, project delivery strategy, major milestones, potential off-ramps, and cost estimates. When the planning effort involves multiple facilities and/or includes other system infrastructure—for example, the collection system—some agencies refer to the facility plan as the "master plan." Modern facility plans can be complex and involve multidisciplinary evaluations including, for example, assessment of the treatment process, infrastructure condition, energy management and electrical systems, instrumentation and automation, asset management, staffing, safety, support facilities, resource recovery management, public outreach, rate studies, and finances. Facility plans have traditionally included two major phases: (a) "plan and benchmark," the development of the basis of planning, and (b) "explore and converge," comprising exploration analysis, selection of alternatives, and development of an implementation plan. This chapter presents the information and evaluations that would generally be included in each of these phases.

The intent is to provide public agencies (i.e., cities, towns, districts, counties) with a basic framework and general guidance to develop their facility plan or master plan. It is important to note that local ordinances and regulations should be reviewed before starting a facility plan to ensure that the regulatory requirements for review and approval of the planning documents are met. Also, in case of special funding applications, it is important that the requirements of the funding agencies are also met.

2.0   PROJECT CHARTERING

At the onset of the planning effort, the planning team and critical stakeholders meet to define the facility plan goals and desired planning outcomes. This group may include an agency's project manager, operations and maintenance (O&M) team, engineering team, regulatory coordinator, planning team, and financing and communication team, along with the facility plan consulting team. As a result, general and specific objectives are documented in a concise form for use and reference during the planning period. The formats of major deliverables are also established, and style guides for narrative and data documentation are created. Communication with external stakeholders such as regulatory agencies is planned, and points of contact are defined.

2.1   Considering Community and Sustainability

Planning goals and specific desired outcomes must focus on the sustainable development of the communities served by the WRRF. For example, resource recovery considerations and expectations must be well-defined at the project chartering level and carried through the planning development. These considerations should address not only the noneconomic environmental benefits, but also the financial effects of improvements to existing and construction of new facilities, including the effects on rates and the agency's overall finances for O&M. Opportunities for O&M savings should be explored as a benefit to the end customer or the community. For example, modern facility plans address not only improvements, expansion, and upgrade needs, but also explore optimization opportunities and use of new technologies for more economical operation of the WRRF.

Furthermore, nuisance impacts on adjacent communities and neighborhoods should be considered from project chartering throughout the planning effort. Facilities that introduce noise, odors, negative visual impacts, traffic, and chemical use and handling should be carefully discussed with community stakeholders. A plan of mitigation measures should be included, and public outreach efforts should be planned to inform the community about the projects at the WRRF.

2.2   Planning Team Engagement

The facility plan sets the stage for the expansion, upgrade, or new construction of a WRRF, and it is important to keep stakeholders involved throughout the process. The plan should reflect stakeholders' needs while maintaining a cost balance through the decision-making process. It is important to include all levels of the agency's organization in the facility planning process. This will establish a spirit of cooperation and buy-in from those individuals who will be charged with making the project a success. In the same light, it is important to match individuals within the facility planning consulting team with their counterparts in the agency. For example, the consultant's lead electrical engineer should establish a connection with the agency's chief electrician, and the team establishing criteria for a new laboratory must consult with the agency's laboratory director and staff. One of the best ways to engage stakeholders is by scheduling informal workshops throughout the planning process to obtain input and feedback. Especially important is the initial workshop or kickoff meeting, where the project vision, goals, critical success factors, constraints, and stakeholder interaction can be started, discussed, and agreed on as a team. Using a facilitator may make this and other key workshops more productive.

2.3   Balancing Electronic Versus Hard-Copy Documentation

Many agencies are adopting a planning cycle that varies from 3 to 5 years. Each cycle includes revisions of the planning drivers, projects, and implementation schedules. To avoid repeating work and to streamline the revisions, agencies are adopting electronic facility planning tools that can vary from simple electronic spreadsheets to complex planning systems incorporating asset management. At the chartering level, stakeholders should define the level of electronic versus hard-copy documentation that will be prepared through the facility planning effort, with tools and software defined and listed. It is important that the agency gives preference to open market software that can be used by its staff or multiple engineering firms in the future, rather than proprietary packages and spreadsheets. Making the facility plan deliverables portable avoids rework and saves on future planning efforts.

2.4   Establishing Synergy with Asset Management

Both facility plans and asset management plans address the managerial aspects of asset ownership, and agencies are investing more on modern asset management tools and approaches. Assets are typically classified hierarchically, for example, in terms of major facilities, major systems, subsystems, single assets, and single components. While asset management plans go deep into the asset hierarchy and address single assets and single components (e.g., valves, pumps, seals, control panels, computers), facility plans maintain a higher level of addressing primarily major systems and subsystems (e.g., preliminary treatment, secondary treatment, aeration basins, clarifiers, aeration systems).

The U.S. Environmental Protection Agency has developed guides for asset management plans, tools, and implementation (U.S. EPA, 2020) that could be helpful when establishing the boundaries between facility and asset management plans. Another useful reference is International Organization for Standardization literature for ISO 55000, which is the international standard for asset management (International Organization for Standardization, 2014).

Establishing boundaries between facility and asset management plans is important to leverage the synergy between the plans, avoid duplicate efforts, promote a single vision, and set expectations regarding the overall facility plan. Boundaries between the facility plan and asset management plan must be communicated with stakeholders to set a uniform level of expectations across the planning team. For example, the maintenance staff should agree that evaluating the replacement of a control valve is within the scope of the asset management program. However, evaluating the replacement of an entire aeration system is within the scope of the facility plan. These boundaries may vary from agency to agency depending on the level of sophistication of the facilities involved. Often, agencies serving small communities do not have a formal asset management program, and the facility plan becomes the guiding document for asset management at all levels of the asset hierarchy. In these cases, careful condition assessment of the facilities is conducted, including assessment of single assets and sometimes single components.

Successful facility plans are dynamic and, like asset management programs, occur in predefined cycles. This programmatic approach is recommended particularly for mid- to large-size agencies. It is important to recognize and establish the different cycles of a planning program. Figure 4.1 illustrates typical planning program cycles that can be considered for implementation. The first cycle is the planning cycle, which is essentially the preparation of the facility plan. This cycle starts with phase 1 (plan and

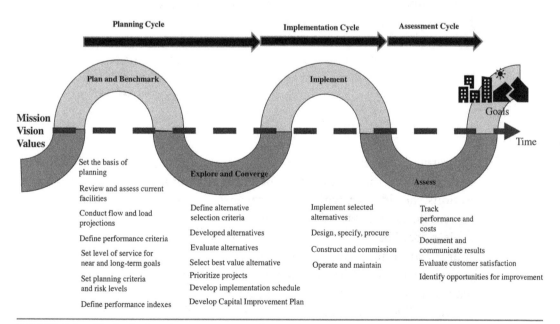

FIGURE 4.1   Typical planning program cycles.

benchmark), where assets are evaluated and improvements are identified. Next is phase 2 (explore and converge), where alternatives are developed and evaluated and projects are selected and packaged for implementation. Then, the implementation cycle handles design, construction, commissioning, and O&M of assets. Finally, the assessment cycle evaluates performance and identifies opportunities for improvements, leading to a new planning cycle.

2.5   Defining Planning Drivers

According to the programmatic approach illustrated in Section 2.4, facility plans are initiated with a plan and benchmark phase. During this phase, it is important to define the planning drivers, which can be related to regulations, growth, or asset rehabilitation/replacement. Chapter 2 describes in detail the definitions of the different planning drivers. The next section focuses on the overall definition and documentation of the basis of planning.

3.0   BASIS OF PLANNING

The basis of planning effort is developed after defining and understanding the planning drivers, but still within the plan and benchmark phase. The goal is to clearly and succinctly document critical conditions and factors that will serve as the basis for the subsequent explore and converge phase. While the planning drivers address the "big picture" considerations, the basis of planning effort focuses on metrics. For example, if a treatment facility foresees a population increase in its service area, the basis of planning effort will investigate and project future wastewater flow and loads to the facility for the planning period. While the basis of planning effort will vary from facility to facility, there are common core elements. The following sections discuss typical considerations for the basis of planning effort.

3.1   Projecting Flows and Loads

Projecting flows and loads are critical to set the requirements for capacity and reliability for the different unit processes in a treatment facility. Projections should start with a comprehensive review and documentation of current flows and loads. Different methods are used for population and flow projections. The most common method uses historical per capita wastewater source data and population projections to estimate future flowrates and organic loads. If historical data are not available, typical per capita flow and load contributions are used and can be found in *Design of Municipal Water Resource Recovery Facilities* (Water Environment Federation & American Society of Civil Engineers, 2017). Facilities that serve large metropolitan areas where flow contributions are affected by commuters may estimate a population equivalent that accounts for population and employment within a given service area.

Given the uncertainties in population projections, some agencies use scenario planning to define a reasonable bracket for planning with low and high population growth scenarios. This approach allows better modularity and can avoid excessive capacity and overspending. In addition, regional WRRFs that provide service to multiple communities, interjurisdictional agencies, and wholesale customers should include projections for the individual agencies they serve.

Industrial and commercial contributions may be accounted for separately as point sources or included in the overall per capita contributions, if the flows and loads are not excessive.

Infiltration and inflow and stormwater contributions should be accounted for in the flow evaluation. The evaluation of these flow contributions may require extensive sewershed evaluation including hydrological analysis, flow monitoring, and modeling. For a detailed description on infiltration and inflow and stormwater contributions, refer to *Design of Municipal Water Resource Recovery Facilities* (WEF & ASCE, 2017).

3.2   Defining Effluent Quality Criteria

Effluent quality criteria are typically defined based on regulatory requirements and overall effluent management requirements. In cases where the regulatory requirements are anticipated to change within the planning period, it is important to develop a pictorial representation of the regulatory roadmap showing the dates where different requirements are triggered. This roadmap may be carefully vetted with the regulatory agency to ensure accuracy of the timeline. It is important to include compliance schedules and regulatory capacity planning and expansion triggers. For example, regulations may require an 80% capacity trigger for expansion planning and a 90% capacity trigger for expansion. This means that at 80% capacity, agencies may be required to submit a WRRF expansion plan or facility plan, while at 90% capacity, agencies may be required to initiate the WRRF expansion.

Water reuse and regional water management practices should be considered. In cases where regional load trading and allocations are allowed, the effluent quality criteria must consider regional trading agreements beyond single discharge requirements. For example, where total maximum daily loads (TMDLs) are established, it is important to estimate the allowable effluent concentration as a function of time. The plot in Figure 4.2 illustrates the allowable effluent concentration for a hypothetical total phosphorus TMDL of 79.4 kg/day (175 lb/day). As the flow increases, the allowable

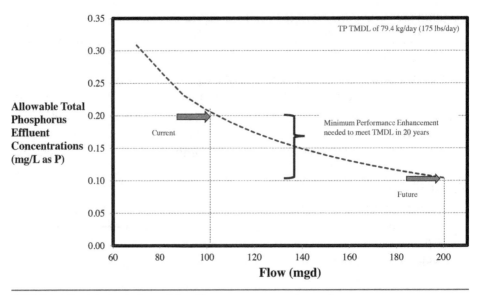

FIGURE 4.2   Typical allowable effluent concentrations plot.

effluent concentration decreases. The future effluent criteria can be estimated by combining the allowable effluent curve with the flow projections for the WRRF.

Facility operation goals should be established for planning and design to provide appropriate safety factors for permit compliance.

3.3   Describing and Assessing Existing Facilities

The description and assessment of the existing facilities is conducted with the participation of O&M staff and engineers. It is important to document the level of service, physical condition, operability, and maintainability of major systems, subsystems, and major assets. It is also important to review and summarize the original design criteria and current process performance, which can be documented in tabular format.

The description and assessment should include a summarized narrative about major systems and subsystems. Many agencies have evolved to electronic platforms for asset management and O&M manuals. For this reason, it is important to verify consistency across documents and avoid duplicating efforts. Typical WRRF planning assessment elements include the following.

3.3.1   Collection System and Influent Interceptors

Influent interceptors and pumping stations adjacent to the WRRF can be included as part of the facility envelope. These are important systems that often get neglected because they may fall between the service area and the facility itself. It is important to describe and document the condition of these facilities as well as capacity and reliability considerations. These investigations must be coordinated with the overall service area planning.

3.3.2 Liquid Stream Processes

Liquid streams may be defined as all treatment systems included in the mainstream process flow sheet of a WRRF. It includes headworks with screening and grit separation, primary treatment, secondary biological treatment, tertiary treatment with filtration and disinfection, and, in some cases, advanced treatment for water reuse applications. The description and assessment of liquid stream processes should include a treatment capacity analysis in addition to the physical assessment of major systems and subsystems. It is fundamental to contrast design load capacity with the current and projected loads. Wastewater strength has changed in many communities as result of water conservation practices. These changes can greatly affect capacity ratings and basis of planning considerations. In addition, some regulatory agencies are adopting load criteria within the permit process.

3.3.3 Solids Stream Processes

Solids streams may be defined as all solids handling and treatment systems that handle sludge and debris produced by the mainstream process flow sheet of a WRRF. They may include screenings and grit handling systems, sludge thickening systems, sludge pretreatment systems, high-strength waste receiving and handling facilities, sludge stabilization systems, solids dewatering and storage systems, and resource recovery facilities (i.e., biogas handling systems, energy recovery facilities, nutrient recovery systems, or other material recovery systems).

The description and assessment of solids stream processes should also include treatment capacity analysis in addition to physical assessment of major systems and subsystems. A solids balance should be developed for current conditions and used as the basis to project solids production for evaluating future needs.

It is important to identify and document current and future treatment requirements in coordination with local and federal regulatory requirements.

3.3.4 Biosolids Management Systems

Biosolids management processes include systems dedicated to beneficial use and final destination of biosolids produced within a WRRF. These processes may include composting and postprocessing processes required for beneficial use and public health protection. They may also include conveyance systems, storage facilities, and land application equipment. Some agencies consider solids streams and biosolids management systems as one major system and develop dedicated biosolids management plans.

From the regulatory standpoint, most facilities will fall within the U.S. EPA Title 40 of the Code of Federal Regulations (CFR) Part 503, often referred to as the "Part 503 Biosolids Rule." Meeting the Part 503 Biosolids Rule requirements will typically drive the technology selection for solids treatment depending on the biosolids classification applicable. Detailed information about specific biosolids classifications under the Part 503 Biosolids Rule can be found in the literature (U.S. EPA, 1994; WEF, 2012).

3.3.5 Sidestream Processes

Sidestream processes include all treatment and handling systems for liquid recycle from solids processing and handling facilities. These processes may include liquid recycle equalization systems, treatment systems, and liquid recycle resource recovery systems. Sidestream processes have become critical for facilities with nutrient limits. They are used to reduce the high-strength liquid recycle nutrient loads in a more economical and compact way. Existing facilities should be described in terms of their treatment goal, design criteria, and capacity. The description and evaluation of existing sidestream processes should carefully include capital and operation expenditures (CAPEX and OPEX) and verify cost benefit though the life cycle of the facilities. New sidestream treatment technologies are available and are being rapidly developed. This provides agencies the opportunity to optimize expenses during the life of the system and leverage new technologies along the way. Incorporation of new sidestream processes should also consider capital as part of the exploration phase of the facility plan, as well as ample technology search.

3.3.6 Odor Control Processes and Air Quality Considerations

Odor control processes include all systems dedicated to management, collection, and treatment of gases produced through the liquid and solids stream processes. These systems can include ventilation fans, collection ductwork, and treatment systems (e.g., chemical scrubbers, biofilters, biotowers).

It is also important to address air quality effects beyond odors. The generation of greenhouse gases (GHGs) is a critical aspect for a sustainable operation. In addition, emissions from generators and combined heat and power systems should be assessed and considered in the overall regulatory review.

3.3.7 Support Facilities

Support facilities include all buildings and infrastructure that support the operation, maintenance, and management of a WRRF. These facilities may include, for example, the administrative building, laboratory, operations center, information and technology center, maintenance shop, fleet services, warehouses, and public outreach auditoriums. They include support systems such as heating, ventilation, and air conditioning (HVAC), potable water, reuse water, compressed air, natural gas, and other ancillary systems. Support facilities should be planned with ample staff participation. These facilities are critical to support team logistics, personal safety, and effective O&M. The participation of architects in the planning team is recommended for proper evaluation of spaces and relevant code compliance. Staff projections can be conducted to evaluate adequacy of existing facilities for long-term services.

3.3.8 Electrical, Instrumentation, and Communication Systems

The facility plan should confirm that adequate electrical, instrumentation, and communication systems infrastructure is in place to accommodate current and future demands. When a facility expansion or upgrade is seen on the planning horizon, the potential additional electrical loads should be estimated and the available power supply verified. Standby power capabilities should also be assessed and planned.

A complete description of the instrumentation and control (I&C) systems should be prepared, including summary of supervisory control and data acquisition platforms and some of the key control strategies. The level of automation for each major system should be defined in collaboration with the WRRF O&M team. Communication systems should be carefully planned and include considerations for cybersecurity.

3.4 Developing a Treatment Capacity Baseline

The capacity analysis is one of the most important elements of a facility plan. It is important to define treatment capacity as a combination of hydraulic capacity and process capacity. For some major systems, the capacity is governed by hydraulic capacity, while others are governed by the process capacity. For example, while the treatment capacity of a headworks process unit is governed mostly by its hydraulic capacity, the capacity of an activated sludge system is typically governed by its process capacity.

3.4.1 Hydraulic Capacity

The hydraulic capacity quantifies the amount of flow that a major process can convey under a set criterion. It is important to coordinate the hydraulic profile with the collection system model to characterize conditions at the influent interceptors tie-in point to the headworks facilities. Typically, the hydraulic capacity criterion is set for peak instantaneous flow and peak hourly flow. While peak instantaneous flow is the highest flow the major process can handle, the peak hourly flow is the maximum peak flow sustained for 1 hour that the major system can accommodate.

To provide a realistic picture of the WRRF's hydraulic capacity, a hydraulic model and profile, as supplements to the fluid balance, should be developed for pumping and open-channel conveyances. A survey of elevations, including hydraulic control points (e.g., weirs, overflows, tops of channels), should be performed to ensure that the control points are represented on the same datum. Once the hydraulic profile is completed, selected water elevations at defined flowrates should be checked by survey. The owner should participate in selecting tools used for the analysis and confirm that the selected software is readily available in the market to facilitate future updates.

3.4.2 Process Capacity

The process capacity is related to the amount of constituent load a major system can treat under a set criterion. Typically, the process capacity criterion is set based on maximum month load conditions that a major system can accommodate. Capacity evaluations are typically conducted through treatment process simulations. Several simulation packages are available in the market and can be selected with the assistance of the facility plan consultant. It is important to evaluate the system with all units in service, but also account for scenarios where units are out of service following the WRRF's established redundancy criteria.

3.5   Considering Site Conditions and Land Use Aspects

Effective site and space planning at early stages of the WRRF service life are critical. Some considerations that must be taken include:

- Land use at the site and within its vicinity. In several major metropolitan areas, residential areas were and are encroaching on WRRFs. This condition creates challenges for the WRRF O&M, such as more stringent odor and noise control, effects on traffic control and safety, and property invasion and intrusion. When possible, a buffer zone within the WRRF property line (i.e., a greenbelt) is highly recommended.

- Flood control and stormwater management. WRRFs are typically at low watershed points and are highly susceptible to flood events. It is important to develop a comprehensive flood control plan for the

WRRF. This effort can be integral to the facility plan or separate. Climate change must be considered with a scenario planning approach that identifies control measures for best and worst cases. It is also critical to consider the effects of new or future facilities on the existing storm drain system.

- Soils and subsurface conditions. A thorough soil and groundwater characterization is recommended at the early stages of planning and should be summarized in the facility plan. WRRFs are typically located on sites that require special foundations or expensive site preparation before construction. Shallow groundwater may also be an issue and must be considered during planning. When projects are triggered for implementation, specific field investigations for soil and subsurface conditions should be conducted to serve as the basis for detailed design, which is discussed in Chapter 7.

- Existing infrastructure is a common challenge in site plans. When practical, storm drains and pipe, duct bank, and utility corridors should be established with initial implementation of the WRRF and planned along the way. The planner must obtain records for site utilities and underground piping for proper siting and placement of new facilities.

3.6   Condition Assessment

Comprehensive condition assessments are typically conducted as part of the asset management plan. Assessment protocols are available in the literature for reference (Water Environment Research Foundation, 2007). Often, a comprehensive condition assessment, including individual assets and components, goes beyond the scope of a facility plan and can incur costs. Therefore, a simplified assessment based on staff interviews and general site visits is commonly adopted, particularly for small WRRFs. In these cases, a matrix can be used to facilitate documentation reference and discussion. It is recommended that this matrix be as simple as possible, with information that is relevant to planning. Condition assessment scores can be included in the matrix and should be based on criteria developed with the participation of O&M staff. It is important to develop an assessment scoring system, or adopt a standard one, to provide metrics for comparison and benchmarking. The scoring system may incorporate risk components with analysis of probability of failure and consequence of failure. An example of a scoring system is presented in Figure 4.3.

The consulting assessment team should include experienced and licensed professionals and involve different engineering disciplines (i.e., mechanical,

Risk Score		Condition Score (Probability of Failure)					Facility Risk	# of Facilities
		1 Very Good	2 Minor Defects	3 Maintenance Required	4 Requires Renewal	5 Unservicable		
Consequence of Failure	C Low Impact	8	10	4	3	1	Low	47
	B Medium Impact	12	6	4	1	0	Medium	7
	A High Impact	4	1	0	1	0	High	1

FIGURE 4.3   Condition assessment scoring system example.

process, structural, electrical, I&C). The team may also include specialty certified professionals like coating specialists, corrosion control experts, and licensed operations professionals.

To facilitate the work of the assessment team, condition assessment forms are used. These forms are typically customized and may vary depending on the assessment.

3.7   Effluent Management Considerations

Effluent management considerations must be incorporated in the facility plan, particularly where water reuse is part of the local and regional water resources portfolio. Exploration of water reuse opportunities must follow federal and state regulations. Typically, states categorize water quality criteria based on the reuse application. Applications may range from nonpotable reuse with less stringent criteria to indirect or direct potable reuse with stringent water quality criteria and multi-barrier treatment trains. Therefore, it is critical that water reuse opportunities balance resource recovery benefits with treatment capital and O&M costs. Some examples of potential effluent management through water reuse include aquifer storage and recovery, groundwater replenishment, saltwater intrusion mitigation, irrigation, industrial cooling systems, WRRF service use, and others. When appropriate, the facility plan can include outside-the-fence distribution systems, storage systems, and groundwater injection systems. Specific considerations for effluent management including water reuse can be found in *Design of Municipal Water Resource Recovery Facilities* (WEF & ASCE, 2017).

3.8   Biosolids Management Considerations

Provided that regulatory requirements are met, biosolids can be beneficially used for agricultural purposes, soil amendment, landfill soil cover, and soil remediation sites. For land application and beneficial agricultural use, it is important to estimate the application rates and land requirements, while also observing nutrient contributions and limits for current and future conditions. It is also important to consider seasonality, weather constraints, transportation/conveyance, storage, and application methods. Infrastructure and workforce are also important factors. Specific considerations for biosolids management can be found in the literature (WEF, 2012).

4.0 ALTERNATIVE EXPLORATION, ANALYSIS, AND SELECTION

In the explore and converge phase of the planning cycle, a full range of alternatives is developed and evaluated to meet needs assessed in the plan and benchmark phase. Innovation and technology should be considered to identify and leverage cost-saving opportunities while matching the WRRF needs. The typical steps of alternative exploration, analysis, and selection are presented in Figure 4.4 and discussed in the following subsections.

4.1 Defining the Selection Approach and Criteria

Alternatives are evaluated and compared based on the approach and criteria developed in collaboration with WRRF staff. This step is highly customized and can vary from facility to facility. Experience has shown that overelaborated selection approaches may lead to time-consuming evaluations and challenging consensus. Therefore, it is important to keep the selection approach and criteria simple, but comprehensive, particularly for small WRRFs.

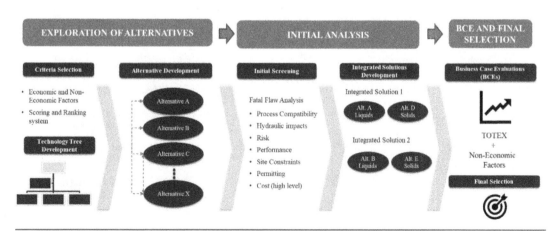

FIGURE 4.4   Alternative exploration, analysis, and selection steps.

One potential evaluation approach for defining selection criteria is outlined below and presented in Table 4.1.

1. Define economic and noneconomic factors.
2. Prioritize criteria based on their relative importance to the project. Assign a relative weight based on a scale of 1 to 10, with 1 being least important and 10 being the most important. Use the paired comparison matrix method to facilitate prioritization.
3. Assign a score, ranging from 1 through 5, with 1 being "poor" and 5 being "excellent," to indicate how well each alternative satisfies each criterion.
4. Apply a weighting factor to the "score" to determine the "points" for each alternative and criterion.
5. Determine the total "points" for each alternative.
6. Rank the options based on the total "points." The alternatives that best satisfy the overall evaluation criteria are top ranked and may be selected for further refinement or, in some cases, the alternative is directly selected for implementation.

Often, the selection process will short-list two or three close candidates. In this case, a more detailed business case evaluation can be conducted to

TABLE 4.1   Example of evaluation criteria.

Economic Factors	Noneconomic Factors
• Capital expenditures	• Safety
• O&M expenditures	• Environmental impacts
• Life cycle cost	• Social impacts
• Resource recovery revenue	• Space requirements
• Monetized GHG emissions	• Track record
	• Reliability and flexibility
	• Complexity and integration with existing facilities
	• Impact on current operations (e.g., power requirements, solids production, recycle streams)
	• Easy expansion and upgrade to accommodate future requirements
	• Constructability and maintenance of operations during construction

better support the decision-making process. This approach is typically used to narrow down selections to a group of alternatives that best meet the evaluation criteria. It offers a cost-effective way to identify suitable alternatives in a systematic manner. The entire array of alternatives can be evaluated using a broad range of criteria to reduce the number of alternatives to a more manageable number. The second screening, with a detailed business case evaluation, could then be used to identify and select the preferred alternative for implementation.

4.2   Developing and Exploring Alternatives

Development and exploration of alternatives must involve the WRRF engineering and O&M team, as well as technical consultants and advisors. A technology review may be conducted with broad exploration of alternatives. The review must focus on the near- and long-term needs identified in the plan and benchmark phase. Technology trees are a useful tool to guide alternative development. The planning team will typically develop a customized technology tree to fit specific needs of each WRRF. This technology tree constitutes an initial universe of possibilities and basis for alternative development. On a collaborative set, the WRRF team and engineering consultant bundle promising technologies and develop implementation alternatives.

4.3   Conducting Initial Screening and Fatal Flaw Analysis

A high-level assessment for fatal flaws is recommended to avoid unnecessary work and focus on feasible alternatives. Some of the checkpoints may include:

- Compatibility with existing processes
- Impact on hydraulic profile
- Technology maturity and risks of early adoption
- Adequacy and performance
- Site considerations
- Permitting and code considerations
- Relative costs

4.4   Developing Integrated Solutions and Short-Listing

Alternatives can be combined as integrated solutions with a bundle of technologies fit to purpose. This means that alternatives for treating liquid streams should be examined in conjunction with their effects on the sludge-processing alternatives. For example, a process for biological nutrient removal (BNR) may be affected by the recycle of nutrient-rich streams from solids dewatering to the head of the facility. Therefore, the implementation of a BNR should be coupled with a sidestream management alternative, which could potentially involve one or multiple technologies.

4.5   Preparing Business Case Evaluations for Short-Listed Solutions

Business case evaluations typically consider the financial, environmental, and social aspects of the alternatives short-listed. The financial evaluation and comparison of alternatives should include both CAPEX and OPEX. Total expenditures (TOTEX = CAPEX + OPEX) for the alternatives can be compared on a present worth, total annual cost, or life cycle cost basis. O&M costs should be carefully estimated based on labor, power, fuel, electricity, chemicals, replacement parts, outside contractors, and other expendables. For equipment with an expected life less than the project life, replacement costs should be included in the analysis. Projected labor needs for the alternatives should be clearly indicated. The issue of increases in skill level and labor required should be dealt with through the planning process, based on the policy direction of the owner's strategic and master plans. The estimated quantities of the items listed above should be multiplied by prevailing unit prices to obtain annual costs, with adjustments for price changes over time. Interest rates for financing, taxes, and inflation should be determined and agreed on.

4.6   Selecting Solutions for Implementation

Each integrated solution is evaluated according to the established criteria results and is typically presented in tabular form for quick comparison. Life cycle costs can be presented in a graphic format considering the planning horizon. It is recommended that integrated solutions be selected following ample participation by stakeholders in a facilitated workshop setting.

5.0   IMPLEMENTATION PLAN

After integrated solutions have been selected for implementation, the associated costs, risks, and priorities must be determined and used to identify project scope and timing. This information informs financial planning efforts to ensure adequate funding is available for projects as they are needed. The combination of project scope, timing, costs, and approach comprise the implementation plan.

Facility improvements can be categorized as capital, asset management, or O&M, depending on their associated drivers and the planned funding source.

- **Capital improvements:** Capital improvements include major unit process changes or equipment improvements that are driven by regulatory, performance (i.e., water quality or O&M), or growth-related drivers. These improvements fundamentally change how the facility operates or the capacity at which it can operate.

- **Asset management improvements:** Asset management improvements involve in-kind replacement of existing equipment or infrastructure based on the assigned useful life of the asset or known deficiencies that need to be addressed to prevent failure.

- **O&M improvements:** O&M improvements are typically small improvements implemented to promote savings in annual O&M expenses. For example, a facility may install a new chlorine analyzer to optimize disinfection dose control, resulting in a reduction in annual chemical consumption.

5.1   Establishing Project Definition, Prioritization, and Sequencing

Improvements selected through the facility planning process can often be grouped together or with existing planned improvements to form projects. Improvements can be grouped based on facility location, timing, and/or when an improvement in one process area affects the solution or approach in another. For example, if solids stabilization improvements are expected to result in increased biogas production such that the existing system capacity is exceeded, the driver for expanding the biogas system is accelerated and should be conducted in conjunction with solids stabilization improvements.

Project priority is determined by considering the drivers behind improvements, which are discussed in Chapter 2. Some drivers could be considered flexible, while others represent hard deadlines. Factors unrelated to treatment can also be considered when prioritizing projects, including the owner's funding availability, values, goals, and initiatives. Because the driver represents the latest date at which a project or improvement needs to be operational, project timing and sequencing must consider total project duration and ensure enough time is allotted for improvements to be implemented.

5.2   Documenting Cost Considerations and Approach

It is recommended that conceptual-level cost estimates be developed following guidelines in the most recent edition of the Association for the

Advancement of Cost Engineering's *Cost Estimate Classification System* (Recommended Practice No. 18R-97) (AACE, 2017). Under these guidelines, estimates are classified as Class 1 through 5 based on the level of project definition, which corresponds to the end use, methodology, expected accuracy range, and level of effort required to prepare the estimate. Facility planning typically uses Class 5 estimates, which are appropriate for concept screening. However, Class 4 or Class 3 estimates may be required, depending on the owner's cost-estimating standards and guidelines.

Costs should include appropriate contingencies based on the available level of detail and probable uncertainties. Specific additional costs for special construction requirements (e.g., site constraints, fast-track situations) must be considered. The accuracy of the cost estimate will vary depending on the level of detail provided in the conceptual design.

An opinion of probable construction cost (OPCC) is an estimation of materials, equipment, and labor required to construct, install, start up, and commission a fully operational system. The OPCC typically includes direct, indirect, and general costs; taxes, bonds, and insurance; and a construction contingency that is proportional to the estimate classification. The OPCC is typically escalated to the midpoint of construction if a total project cost is not being calculated. The total project cost estimate (TPCE) is derived by applying factors for an additional estimate contingency required by the owner (0% to 30%) and contingency for engineering and administrative costs (15% to 25%). The TPCE is then escalated to the midpoint of construction.

Planning-level cost estimates can also be coupled with a risk analysis to incorporate uncertainty in the evaluation. There are several tools available for assessing financial risk. Typically, cost inputs are coupled with a probability distribution and confidence interval, resulting in a range of possible outcomes. Minimum and maximum values can be calculated by including a skew, which represents the degree of uncertainty (i.e., a positive or negative skew gives the likelihood of the actual value falling to the right or left of the average).

5.3   Developing a Facility Roadmap

A roadmap can be developed after improvements have been prioritized. This roadmap visually depicts how identified drivers influence the timing of improvements. It should incorporate recommended studies and evaluations that may influence the solutions implemented under subsequent projects. Roadmaps also identify alternative approaches and "off-ramps." Alternative approaches represent technologies or solutions that were determined to be feasible in the screening process but were not recommended as the optimal solution at the time of the evaluation. Off-ramps represent complete deviations from the planned approach. An example of a roadmap is shown in Figure 4.5.

5.4   Preparing the Capital Improvement Plan

A capital improvement plan (CIP) should clearly define projects proposed by the facility plan and its associated costs, project initiation date, and duration. The CIP contains a summarized expenditure schedule, including the anticipated spending for each project in each year over the planning period. It is helpful to group improvements by area or major facility, as certain personnel may only be interested in some of the improvements. An example of a CIP is provided in Table 4.2.

5.5   Identifying Project Delivery Methods

While the facility plan represents the first step of a traditional design–bid–construct construction mechanism, there has been a trend toward other delivery methods, such as design–build, construction manager/general contractor, construction management at risk (CMAR), and design–build–operate. Once the recommended projects have been selected, alternative delivery methods can be explored. The owner's and state procurement rules will affect the type of delivery options that are available. An analysis of the effects of different project delivery methods on schedule, project cost, quality

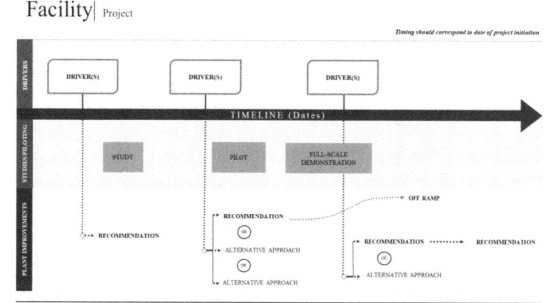

FIGURE 4.5   Example of a roadmap template.

TABLE 4.2   Example of a capital improvement plan expenditure schedule.

Capital Improvement Plan Expenditure Schedule

Category	Projects	OPCC ($)	Duration (Years)	Start Year	OPCC at mid point at 3.0% int. ($)	Engineering and Admin. (Typ. 25%) ($)	TPCE at mid point of const. (+10% contingency) ($)	TPCE at mid point of construction ($MM)	2020	2021	2022	2023	2024	2025	2026	2027	2028	2029	2030
Preliminary / Primary Treatment	Enhanced Primary Treatment Study	240,000	1	2020	240,000	-	240,000	0.24	0.12	0.12									
	Headworks Improvements	4,270,000	2	2022	4,960,000	1,240,000	7,000,000	7.00			2.80	4.20							
	Nutrient Removal Study	500,000	2	2021	-	-	500,000	0.50		0.25	0.25								
Secondary Treatment and Disinfection	Blower Replacement	11,610,000	3	2023	14,070,000	3,520,000	20,000,000	20.00				2.00	9.00	9.00					
	UV Disinfection Retrofit	8,140,000	4	2023	10,020,000	2,510,000	14,000,000	14.00				1.40	4.30	4.90	2.30				
	Full-plant BNR Conversion	25,010,000	5	2023	31,220,000	4,690,000	40,000,000	40.00				8.40	3.60	12.00	12.00	12.00			
Biosolids	Thermal Hydrolysis Pilot	500,000	1	2021	-	-	500,000	0.50		0.50									
	Digester Mixer Replacement	2,770,000	2	2022	3,220,000	810,000	5,000,000	5.00			0.50	2.00	2.50						
	New Thickening Centrifuge Facility	12,700,000	3	2025	16,330,000	4,090,000	25,000,000	25.00						2.50	8.75	8.75	5.00		
Energy	Energy Management Evaluation	500,000	1	2021	560,000	140,000	1,000,000	0.50		0.25	0.25								
	Electrical Gear Replacement	2,700,000	3	2022	3,180,000	800,000	5,000,000	5.00			0.50	2.25	2.25						
Buildings	Admin Building Improvements	2,000,000	2	2029	2,860,000		4,000,000	1.00										1.00	1.00
	New Regional Laboratory	9,990,000	3	2025	12,850,000	3,220,000	18,000,000	18.00						1.80	6.30	6.30	3.60		
	Facility HVAC Upgrades	1,400,000	4	2025	1,830,000	280,000	3,000,000	3.00						0.30	1.05	1.05	0.60		
Total		82,330,000						139.74											

Annual Expenditure

Study/Evaluation Phase
Design Phase
Construction Phase

OPCC - Opinion of Probable Construction Cost
TPCE - Total Project Cost Estimate

of construction, and owner control should be provided to assist the owner in making an informed decision. Alternative delivery methods are discussed in Chapter 5 of this manual.

5.6   Preparing Rate Study and Funding Evaluations

Because the cost of constructing and operating the facility is affected by the financing plan, a capital and cash-flow schedule should be considered. The cash flow should show the level of funding and low-interest loans anticipated (and grants, if available); use of reserve funding; and the plan for issuance and repayment of the bonds that will be issued by the owner. The assumed interest rates, escalation factors, and schedule for the repayment of bonds and loans should be clearly identified. User charges should be projected for the life of the project to show how the cost for continued project O&M will be recovered. The user charges can then be compared to income levels to demonstrate project affordability. When appropriate, the project can be divided into specific phases that match the availability of funding balanced with capacity requirements to construct the new facilities.

5.7   Developing a Public Outreach Plan

Regulatory review and public participation and involvement are major elements of a successful facility plan. Because residents living in the neighborhood of the WRRF will be directly affected by the planned improvement(s), they should be included in the planning process as early as possible to obtain their input and garner their support. In some cases, a public session can be held at the beginning of the planning process to identify those issues of concern to the public. Often a small thing, such as a few additional parking spaces, improved storm drainage, a park, or alternative truck route can go a long way in obtaining public support. Effective tree planting can reduce odors and increase the aesthetics of the perimeter better than simply fencing (i.e., "out of sight, out of mind"). The public can become involved through various types of communication, including informational meetings, flyers, and newsletters. If appropriate, a citizens' advisory committee should be established for large projects and given a budget and autonomy to provide an independent review and opinions. A plan and procedures should be developed for dealing with public complaints, accidents, and insurance claims. A formal public hearing should be held once the recommended plan is selected to obtain public input, and a formal response issued to address any comments that are received. Required regulatory reviews should be scheduled, discussed proactively with those review staff, and submitted in a timely manner. The engineer and owner need to document the response to regulatory comments and modify the plan as necessary to obtain final approval before proceeding with the detailed design.

6.0 REFERENCES

Association for the Advancement of Cost Engineering. (2020). *Cost estimate classification system—As applied in engineering, procurement, and construction for the process industries* (AACE Recommended Practice No. 18R-97). ACCE International.

International Organization for Standardization. (2014). *ISO 55000: 2014: Asset management: Overview, principles, and terminology.* https://www.iso.org/standard/55088.html

U.S. Environmental Protection Agency. (1994). *A plain English guide to the EPA Part 503 Biosolids Rule.* EPA/832/R-93/003. https://www.epa.gov/sites/production/files/2018-12/documents/plain-english-guide-part503-biosolids-rule.pdf

U.S. Environment Protection Agency. (2020). *Reference guide for asset management tools.* https://www.epa.gov/sites/production/files/2020-06/documents/reference_guide_for_asset_management_tools_2020.pdf

Water Environment Federation. (2012). *Solids processing design and management.* McGraw-Hill.

Water Environment Federation. (2017). *The water reuse roadmap.* McGraw-Hill.

Water Environment Federation & American Society of Civil Engineers. (2017). *Design of municipal water resource recovery facilities* (6th ed., WEF Manual of Practice No. 8). Water Environment Federation.

Water Environment Research Foundation & American Water Works Association. (2007). *Condition assessment strategies and protocols for water and wastewater utility assets.* Water Environment Research Foundation.

5

Project Delivery Systems

Marie S. Burbano, PhD, PE, BCEE; Rebecca Dahdah, PE; & Brian Shell

1.0   INTRODUCTION

Early in the process of initiating a major water resource recovery facility (WRRF) upgrade, an owner must consider how the project will ultimately be delivered. As water quality professionals, we are all focused on addressing the design and operations needs of the treatment process. However, we need to carefully consider how the final project will be delivered; that is, what will be the roles, responsibilities, and relationships of the various parties involved in design and construction of the final project. This chapter will serve as an introduction to the various project delivery systems an owner may consider for any size or type of facility upgrade or expansion project.

Most public works projects in the United States, including WRRFs, are constructed using the design–bid–build delivery system. There are alternatives to this project delivery method that should be evaluated and considered. This chapter provides a brief history of construction delivery methods and discusses a few of the more common methods currently applied in this industry. The advantages and disadvantages of these methods will also be presented. This information can be used to determine, early in the project, what delivery method(s) should be used. A more thorough and detailed discussion of these delivery methods, including industry standard forms, is available from the suggested readings section listed at the end of this chapter.

The delivery methods that will be discussed in this chapter include the traditional design–bid–build, the design–build, construction manager at risk (CMAR), and public–private partnership. Section 4 on design–build delivery will also discuss fixed-price design–build, progressive design–build (PDB), and design–build–operate (DBO).

2.0 HISTORY AND OVERVIEW OF PROJECT DELIVERY SYSTEMS

Throughout history, a single person was often responsible for the design and construction of a major project. Referred to as a "master builder," their duties ranged from overseeing a design team and a construction team to involvement in the smallest detail of the materials, labor, and equipment used to complete the project. Many relatively recent major projects, such as the Hoover Dam, were essentially constructed this way.

More recently, U.S. public works projects, including WRRFs, were designed and constructed using the design–bid–build or "conventional" delivery method. The separation of the design and construction functions is intended, in part, to provide a system of checks and balances throughout the delivery of the project. It is characterized by a design, prepared by the municipal entity or a consultant (the design engineer), being completed and then publicly bid.

Alternative delivery systems can be simply defined as those that do not follow or are "alternatives" to the conventional process. There are numerous alternative delivery systems and multiple variations of each approach such that a detailed discussion of these could fill an entire book. For the purposes of this chapter, a few of the more commonly used alternative delivery systems in the municipal wastewater treatment industry will be presented. The design–build alternative delivery method has become more common in the last several decades.

Owners or wastewater professionals desiring to explore the variety of possible delivery systems should consult with their engineering professional or attorney, or review some of the suggested readings presented at the end of this chapter. Numerous professional organizations, including the Engineers Joint Contract Documents Committee, the Design–Build Institute of America, Associated General Contractors, American Institute of Architects, and the Construction Management Association of America have published standard documents, manuals, and guides on conventional and alternative project delivery methods.

2.1 The Professional Separation of Design from Construction

During the planning of phase and selection of a project delivery system for a new facility, it is imperative for the owner to recognize the close relationship between design and construction. To ensure project success, the design and construction professionals should openly discuss project expectations and design and construction team relationships before the project starts. This should include defining the division of responsibilities among the design and construction professionals, project milestones, and anticipated deliverables through the duration of the project. These professionals should be rewarded for the responsibilities, risks, and tasks they are to assume. The outcome of these discussions should be formalized in the contract between the parties. The owner should be familiar with the ethics requirements for the jurisdiction where the project will be built and where the work will be performed. The owner should also develop a clear understanding of the requirements and decide who will be responsible for signing and sealing documents in design and construction. Early in the planning process, legal advice should be requested as needed to address questions about developing contracts, ensuring proper ethics, and defining design and construction professional responsibilities.

It is essential that the owner take responsibility for providing adequate time and funding, including approval of scheduling, to allow the design and construction professionals to perform their work in a satisfactory manner. For construction planning, identifying required resources, activities, and expected durations and deadlines will allow for the design to become real and link expectations to the design. In both design and construction, numerous operational tasks must be performed with consideration for precedence and other relationships among the different tasks. Design and construction of a facility must satisfy the conditions unique to a specific site.

The owner should decide on the course of design and approval of shop drawings. This can be done by establishing a clear procedure for the shop drawing process and a thorough designation of responsibility in all the project documents. The owner must assign responsibility for shop drawing review, approval, and stamping in the contracts between the owner and designers and in contracts among the design team. Specifically, the owner should establish adequate flexibility in contract documents to identify shop drawing requirements down the line and provide for necessary submittals, reviews, and approvals.

2.2   The Spearin Doctrine

Before the 1918 U.S. Supreme Court case *United States v. Spearin*, all construction risk was assumed by the project's contractor, unless contracts were modified to say otherwise or the contractor was unable to perform the work because of an "Act of God or Nature." The Spearin Doctrine changed that when the Supreme Court ruled:

> [If] the contractor is bound to build according to plans and specifications prepared by the owner, the contractor will not be responsible for the consequences of defects in the plans and specifications. . . . This responsibility of the owner is not overcome by the usual

clauses requiring builders to visit the site, to check the plans, and to inform themselves of the requirements of the work. (*United States v. Spearin*, 1918)

Therefore, during the preparation of a contract, the owner may choose to reference the Spearin Doctrine to allow the contractor to reasonably assume that the contract documents are adequately designed to protect the contractor in the case that the design proves to be erroneous, more costly, more difficult, or requires more time than contractually anticipated. Furthermore, noting the Spearin Doctrine legally implies that the owner warrants that the plans and specifications are accurate and suitable for their intended use. However, the contractor also has responsibility to convey to the owner any errors in the drawings or specifications, which may not permit compliance or proper completion of a project.

2.3   Financing of Public Works

There are many ways to fund WRRF projects, such as bonds, long-term and short-term financing, and federal or state grants. For smaller projects, owners may wish to use reserved cash. However, maintaining capital reserves is imperative for addressing expenses tied to unforeseen emergencies and ensuring stability under normal operation. Bonds often require a minimum fixed cost for issuance, which may lead owners to consider using cash instead. Looking for options that can provide flexibility in terms of how funds are spent is key when choosing a bond. Flexibility that permits the temporary reimbursement of expenses can, in turn, allow for the money to be used to pay for smaller expenditures later. Financial advisors, such as a bond counsel, should be consulted about the timing, restrictions, regulations, reimbursement, refinancing, and available tax-exempt financing options.

When deciding to finance a project by incurring debt, determining whether to use long-term or short-term debt can be clarified by evaluating interest rates, the useful life of the proposed infrastructure, the terms of the debt, risks, liquidity benefits, and effect on ratepayers and credit ratings. Developing a financial asset and liability management plan can help find the proper mix of investment types.

The U.S. Environmental Protection Agency (U.S. EPA) and the U.S. Department of Agriculture (USDA) have programs that can help fund water and environmental projects. The EPA's Clean Water State Revolving Fund program regularly funds a range of water-related projects. The program allows for states to fund loans; the states can then add a 20% match and are then able to provide a low-interest-rate loan to communities for high-priority water quality projects. The revolving fund allows for new loans to be provided to other recipients as the loan is paid. The USDA's Rural Utilities

Service has more than $4 billion in direct loan funds and $1 billion in grant funds available for water and wastewater projects in rural communities having less than 10,000 people. The USDA's Water and Environmental Programs offers long-term, low-interest loans and grants.

2.4   Phase Sequencing

The main phases of a project are planning, design, award, construction, and completion. During planning, the scope of a project as well as the expectations and delivery method are determined. The owner will determine the budget, schedule, and required team that will carry out the project. The design phase can start after planning, allowing for the owner to engage and approve the progress of the drawings and specifications for a project. The award phase assigns a contractor based on a submitted bid and qualifications. The construction phase will involve trade-specific contractors to follow the requirements outlined in the contract documents. In the completion phase, inspection, testing, and approval of all installed equipment will be performed.

2.5   Factors Influencing Choice of Project Delivery Method

Deciding on a project delivery method for a new project will be a commitment to a comprehensive process that will set the pace of the planning phase and continue through design and construction phases. There is the possibility of tailoring a delivery system to a project; however, each of these delivery methods establishes different relationships among the parties involved and implies different levels of risk. Factors the owner should consider when deciding which delivery method is best for a project include:

- Level of owner control
- Owner relationship with selected team
- Project budget
- Project schedule
- Risk to the owner

2.6   Contracting Strategies

Owners should become familiar with choices for establishing and selecting their preferred contractual relationships, owner influence, and project costs, which will also be tied to the project's contract. The terms that the owner of the WRRF and the engineer, design–builder , and contractor agree to will determine the work atmosphere and performance of the project team and will always be tied to the selected project delivery method. The owner may consider using cost-plus-fee, guaranteed maximum price (GMP), lump-sum (or fixed-price), or target price/unit price as possible formats for the project's contract. Contract development requires the owner to consider a realistic budget, schedule, expectations, communication requirements, design process, and risk assessment, with allocation of risks to the appropriate parties and a recognition of the level of expertise within the owner's organization.

3.0   TRADITIONAL DESIGN–BID–BUILD DELIVERY

The design–bid–build or conventional project delivery method has been used in the United States for over a century. It is the traditional, most commonly used project delivery method for public works projects and is allowed by all known local and state governments. As such, it is also a familiar process for owners, consultants, material and equipment suppliers, and contractors. This is an immediate advantage for this delivery method because of each party's level of comfort and understanding of their roles and responsibilities.

In the design–bid–build project delivery method, the owner either completes the design of a project with the owner's engineering staff or retains the services of a consulting engineer to perform project design. If a consulting engineer is retained, the engineer's responsibility, authority, and other requirements are defined in the contract between the owner and the engineer. This is often the same consultant used for any assistance that was provided during the earlier planning phases of the project. The design engineer is responsible for preparing contract documents that enable construction contractors to understand the requirements of those contract documents, and bid fairly and equally on the contract.

Following completion of the contract documents, construction contractors are invited to bid on the project. Generally, the contract is then awarded to the lowest responsible, responsive bidder (contractor). Sometimes, contractors are prequalified before bidding. The contractor is then retained under a separate contract with the owner. This relationship is shown in Figure 5.1, which is a commonly used diagram. Frequently, the design engineer is retained to assist the owner with a variety of services during construction, such as shop drawing review and construction inspection. It is important to note that, in the design–bid–build project delivery method, there is no contractual relationship between the design engineer and the contractor. It is imperative that the contract documents define the roles, responsibilities, and relationships among all parties to minimize problems during construction. Table 5.1 summarizes the typical responsibilities of each party in this process.

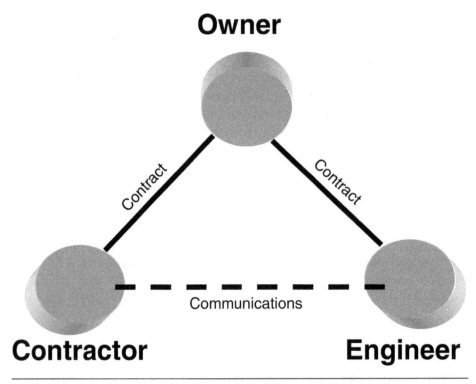

FIGURE 5.1   Owner, engineer, and contractor relationship.

3.1   Sequencing of Tasks

In the traditional design–bid build delivery method, the sequencing of tasks starts with the planning or predesign phase; a design team is defined and then the design phase can start. The design phase can be split up into different milestones, with deliverables submitted to the owner for approval of design progress. The deliverables will be reviewed and approved by the owner for the final development of drawings and specifications, which will form the contract documents. The contract documents must be reviewed and approved by corresponding regulatory agencies, which will issue permits for construction. The contractor can be selected during the preconstruction phase after the completion of the contract documents by the design team. Contractors can submit bids for evaluation by the owner. The owner should review qualifications and will then submit documents for final award and the start of the construction. Once all aspects of the project are constructed, the contractor will need to obtain a certificate of occupancy and achieve substantial completion through approved inspections performed by the design team. With the designation of final completion, the project can be used for its designed operation.

TABLE 5.1   Design–bid–build project team and responsibilities.

Team Members	Responsibilities
Owner	• Determining the goals and requirements for the project
	• Acquiring a usable site for the intended purposes
	• Financing the project
	• Directing both engineer and contractor
	• Resolving issues between the engineer and contractor
Design Engineer	• Predesign
	• Assisting the owner in determining goals and requirements
	• Developing the project's design
	• Coordinating consultants
	• Processing entitlements related to design responsibilities, such as planning approvals and zoning variances
	• Processing building permits
	• Ensuring regulatory and code compliance
	• Preparing construction documents (i.e., drawings and specifications)
	• Estimating the probable construction cost
	• Assisting with construction bidding or negotiation
	• Reviewing and approving shop drawings
	• Administering the contract for construction
Contractor	• Committing to the cost of construction
	• Obtaining entitlements related to construction, such as building and encroachment permits
	• Preparing shop drawings and other submittals necessary to accomplish the work
	• Providing the methods and means of construction
	• Coordinating subcontractors
	• Establishing and maintaining the construction schedule
	• Ensuring job-site safety
	• Fulfilling the requirements of the construction documents
	• Guaranteeing the quality of construction
	• Correcting deficiencies covered by the guarantee

3.2   Project Team and Primary Functional Relationships

The project team for design–bid–build delivery includes the owner, the design engineer, and the contractor. Refer to Table 5.1 for the responsibility of each team member in a design–bid–build project.

The functional relationship of a design–bid–build project team could result in the following problems:

- Adversarial relationships among all parties to the project
- Disputes because of unrealistic expectations in the contract documents and the requirement to accept the lowest bid.
- Material and equipment selection by the construction contractor potentially causing problems because of substitutions or "equals," or choosing excessively low-cost items

Both the owner and the design engineer should perform due diligence to ensure that the design–bid–build project delivery method is appropriate.

3.3   Uses

The design–bid–build project delivery method is the most commonly used approach for the design and construction of public works projects. In many cases, this is simply because of past practice or requirements of local or state law. It is often specifically chosen when options are available and when the owner wants to ensure maximum input regarding the details of the design, has adequate time for this sequential process, and desires the benefits of an independent engineer representing owner interests during design and construction. For municipal WRRF projects, the bids are often required to be a single, lump-sum figure, simplifying bid analysis and award. Unit price contracting, where the contractor bids a unit price for several individual items, is more common in "linear" projects such as roads, sewers, and water mains.

Based on history, precedent, and some of the advantages of the design–bid–build delivery method, it is expected to continue to be the most commonly used approach for years to come.

3.4   Procurement

The procurement phase starts with the completion of the design phase in the design–bid–build delivery method. The construction documents are distributed to contractors who prepare bids to complete the proposed project. The various contractors will prepare bids, which will also include bids from subcontractors. During the bidding period, any questions or clarifications that the contractor has will be resolved with the issuances of addenda. Once the bids close, the lowest bidder will be responsible for performing the work for public sector projects. However, for private sector projects, the owner may decide to evaluate all the bids to determine the winning bid. If the bids from the various contractors are not similar in cost to one another or to the owner's or engineer's estimate, then the owner may decide to cancel the bid and rebid the project, cancel the project, or proceed with value engineering. Value engineering will review construction documents to determine if there are any means of reducing the cost of construction.

3.5   Advantages and Disadvantages

The design–bid–build approach offers the following advantages:

- High level of comfort of all parties because the method is commonly understood.
- Roles and responsibilities are clearly defined and well established.
- Provides a system of checks and balances because the engineer and the contractor are independently retained by the owner.
- Owners can more actively control, direct, and participate in the design process and the resulting final designed project.
- Sequence of events do not overlap, simplifying attention to time for each step.
- Legal history, precedent, and case law help clarify areas of potential dispute.
- Insurance and bonding programs are well defined.
- Complies with all known local and state laws.
- Applicable for any size or scale of project.

The disadvantages of the design–bid–build delivery method include the following:

- Lengthy process because each step (design, bid, build) is sequential.
- Construction cost is not established until after the design is fully completed. If the cost does not meet owner expectations, delays result from redesign, value engineering, and rebidding.
- Owner is often put in the position of managing disputes between the contractor and engineer.
- Excessive and sometimes seemingly contradictory legal precedent exists.

- Owner, under contract with the contractor, will be party to any claim of adequacy or constructability of the contract documents.
- Engineer may not have access to or benefit from information from contractors and subcontractors.

3.6   Risks and Limitations

The traditional design–bid–build delivery method does not develop any contractual obligation between the designer and the contractor, leading to the owner bearing the majority of the responsibilities associated with the completeness of the contract documents.

The design–build–delivery method requires linear phasing of a project, which limits the dynamics between the design and construction. Any changes required in the design or schedule will need to be absorbed through change orders, which may be costly to the owner.

4.0   DESIGN–BUILD DELIVERY

The design–build method of project delivery is primarily characterized by the owner executing a single contract with one entity. In the 1990s, the design–build approach gained significant popularity, including in the municipal wastewater industry. The growth of this project delivery method has greatly accelerated in the 21st century, making this one of the most significant trends in the construction industry. Its use is even more common in private industry.

Because of the rapid rise in the popularity of this delivery method, numerous engineering, construction, and related organizations have prepared guidance documents, standard form agreements, and other items to assist the owner and design–builder in the effort to define their roles and responsibilities. These organizations include:

- Water Design–Build Council
- Engineers Joint Contract Documents Committee—sponsored by the American Society of Civil Engineers, National Society of Professional Engineers, and Construction Specifications Institute
- American Institute of Architects
- Design–Build Institute of America
- Associated General Contractors of America
- Construction Management Association of America

The reader is referred to these organizations for information on the documents they offer.

The three types of design-build delivery discussed in this section are: (1) fixed-price design–build, progressive design–build, and design–build–operate.

4.1   Fixed-Price Design–Build Delivery

4.1.1   Sequencing of Tasks, Contracting Strategy, and Procurement

The most common approach to the design–build–delivery method first requires the owner to adequately define the project so that the design–builder has enough information to prepare the bid and negotiate. This is generally done through preparation of design contract documents, which can vary significantly in their level of detail and completion. These are sometimes referred to as tender documents, bridging documents, or preliminary design documents. Frequently, they describe the scope, materials, site data, performance criteria, and requirements for the equipment and systems to be used. The owner may retain a consultant to prepare these documents and protect the owner's interests throughout the remainder of the design–build process, including oversight of the design–builder during construction.

Once the project has been defined, the owner procures the design–builder. There are numerous methods for this. One of the most common is a competitive proposal process, where prospective design–builders submit a proposed fee for a finished project. Sometimes, this is done following a separate step during which prospective design–builders are prequalified for bidding. Often, these proposals include qualifications information, details on the proposed final design, estimated operating costs, and other pertinent information. The owner benefits from this effort by the design–builder by observing some of the innovation that the owner is considering for the project. The owner can also benefit by selecting not just the low construction cost, but also selecting a proposal that offers greater reliability, lower operating costs, better aesthetics, or other factors that may have equal or greater importance to the owner for that particular project. It is important, however, to ensure that performance requirements are met through testing, demonstration, or other similar requirements. Other techniques for procuring the design–builder include hard bidding, evaluated bidding (considering construction and operating costs with some kind of system to determine the best bid), and negotiated contracts. Any of these methods can be used through open competition among the design–builders, acceptance of proposals from only prequalified design–builders, or preselection of a few design–builders from which to accept proposals.

4.1.2   Project Team and Primary Functional Relationships

Another aspect of the design–build process of which the owner needs to be aware is the organizational structure of the design–builder. There are a number of firms in the wastewater industry with the qualifications, experience, staff, and interest to complete the design–build project primarily in-house. These tend to be larger construction contractors or engineering firms that have both construction and engineering capability. Often, a contractor and an engineering consultant, and other contributors to the design and construction process, will team together to submit a proposal for a design–build project. These teams can be led by either the contractor or the engineer. Both approaches are common and have inherent advantages and disadvantages. The reader is referred to the suggested readings section at the end of this chapter for a more detailed discussion of this topic.

Once selected, the design–builder must complete the design and then the construction under a single contract with the owner. The owner has significant discretion in the required content of the final "design" documents to be provided by the design–builder. Some documents require little additional design from the design–builder other than providing record drawings. Other documents require a significant quantity of detailed final design drawings, including such items as wiring diagrams, process and instrumentation diagrams, and valve schedules. Any possible savings resulting from a less complete design effort can be significantly reduced or lost if the design–builder is required to provide a more complete design. However, the advantages of single-source responsibility and opportunity for innovation remain.

One factor that the design–build delivery method has in common with the design–bid–build process is the need for well-defined roles and responsibilities for both parties in the contract between the owner and the design–builder. Generally, the design–builder has much more flexibility in the performance of the project than with the design–bid–build approach.

The owner is responsible for activities such as:

- Determining the goals and requirements for the project, sometimes to a high degree of specificity
- Acquiring a usable site for the project
- Financing the project
- Preparing the materials for the design–build entity's selection
- Directing the design–build team

The design–build team is responsible for design activities such as:

- Developing the design for the project within budgetary commitments
- Processing entitlements related to design responsibilities, such as planning approvals and zoning variances
- Ensuring regulatory and code compliance

- Preparing estimates of the probable construction costs
- Preparing construction documents

The design–build team is also responsible for construction activities such as:
- Guaranteeing the actual cost of construction
- Obtaining entitlements related to construction, such as building and encroachment permits
- Maintaining the construction schedule
- Preparing shop drawings and other documents necessary to accomplish the work
- Coordinating the bids and work of subcontractors and prime trades
- Ensuring job-site safety
- Providing methods and means of construction
- Fulfilling the requirements of the construction documents
- Guaranteeing the quality of the construction
- Correcting any deficiencies covered by the guarantee

4.1.3 Uses

There are several reasons an owner would consider the use of the design–build project delivery approach. One of the most common is the desire to avoid the adversarial relationship and associated coordination time, disputes, and litigation frequently associated with the design–bid–build process. It seems that this alone often drives owners to consider design–build. An owner that needs to define the project cost early would lead to consideration of design–build, although the project must be defined well enough to establish a fair, firm price. Another major driver for consideration of the design–build process is the potential time savings. By not having to take the design to full completion, and saving time from the occasionally lengthy bidding process under design–bid–build, owners can shorten the overall project schedule, possibly significantly.

4.1.4 Advantages and Disadvantages

The following are the advantages to the design–build project delivery method:
- The greater the level of detail in these preliminary documents, the more control the owner has over the design. However, the greater the detail, the more is spent—in both time and money—on the preliminary design.

- Design–build allows innovation to occur during the completion of the design and construction. The more detail provided in the preliminary design, the less opportunity there is for the design–builder to innovate.
- The method can save money by not paying for the engineer to complete the documents to the same level of detail as would be required for design–bid–build delivery.

A significant part of the engineer's effort in the design–bid–build process is to provide the design detail required to adequately bid the project and to supply information for the competitive bid process.

Disadvantages of the design–build project delivery method include the following:

- Careful planning and procurement of the design–builder are required for project success
- Lack of familiarity with the process by owners and the public
- Institutional barriers such as local and state public works procurement , or licensing requirements for the engineer of record or for the construction contractor
- Owner loss of control, especially with poorly defined requirements for the design–builder

4.2   Progressive Design–Build Delivery

In a PDB procurement, a design–builder is selected based primarily on qualifications and, where local practice requires it, limited pricing information. As the design–builder develops the design, a construction cost estimate is progressively developed. Once the design is well advanced (beyond 60% and often up to 90%), a GMP is defined for approval by the owner.

If the design–builder and the owner cannot reach agreement on an acceptable GMP or lump sum, the owner can use the completed design as the basis for procuring a hard construction bid. In this case, an "off-ramp" occurs and the project becomes more like a contract design–bid–build, which may affect design ownership.

4.2.1   Uses

The PDB method is often preferred when a project lacks definition or final permitting, or when an owner prefers to remain involved in the design process while leveraging the schedule, collaboration, and contractual advantages provided by a design–build approach. This model is also valuable when regulatory permitting requires well-developed design solutions, or when an owner believes that it can reduce cost by participating in design decisions and by managing risk progressively through the project definition phase.

Owners do not generally use the PDB procurement method when a project's definition is well advanced before the procurement or when a lump-sum construction price is preferred (or required) to select a design–builder.

4.2.2 Advantages and Disadvantages

Advantages of the PDB project delivery method include the following:

- Increased control over project design, construction, and O&M life cycle costs because final contract is not signed until a large portion of the design is complete
- Single, straightforward, and inexpensive procurement process can be completed in a short time frame
- Increased marketplace interest as a result of relatively low proposal preparation cost
- Allows selection of designer and contractor based on past performance, qualifications, and ability to work as a single-entity team with aligned interests for project success
- Provides progressively accurate contractor estimates of total project costs from earliest point in project through GMP definition
- Provides maximum opportunity for designer, contractor, and owner collaboration to define scope, meet schedule and budget, and tailor subcontracting plan
- Provides an "off-ramp" to hard-bid construction if GMP is not competitive or cannot be agreed upon
- No contractor-initiated change orders
- Requires little or no design to be completed by owner in advance of procurement, and provides maximum flexibility in a final determination of project viability for economic and noneconomic factors
- Provides a performance risk transfer mechanism that can be implemented in conjunction with long-term operations commitments
- Single contract and point of contact with owner

Disadvantages of the progressive design–build project delivery method include the following:

- Requires selection based on fee; full construction cost is not known at the time of initial contract

- Existing project design investment may not be of value or use to design–builder
- May not be as fast to deliver as other design–build methods because of potential for extended design/estimate development period, including involvement of numerous stakeholders in the design process
- May not be perceived as being "competitive" for construction pricing
- Requires significant owner and staff involvement and resources during design
- May limit local/small subconsultant participation because of at-risk nature of the work

4.3   Design–Build–Operate

In this approach, the owner uses the design–build approach for the design and construction of the project. However, on completion of the construction, the firm with whom the owner has contracted is also responsible for the operation of the facility for a specified period of time. The time for operations responsibility can be anywhere from 1 year to more than 20 years. The selection of the firm responsible for the DBO contract commonly includes an evaluation of both the construction cost and the operations cost. As with the design–build delivery method, it is critical for the owner to have adequately defined the project requirements in their DBO procurement documents.

There are several other approaches that incorporate the ability of parties other than the owner to be responsible for the project, including financial responsibilities. These are covered in Section 6 on public–private partnerships delivery.

4.3.1   Uses

There are a number of reasons why an owner might consider including an element of operations in a project delivery method. In some cases, the scope and magnitude of the new project are beyond the capability of a municipality's current staff or its ability to retain competent staff in a timely fashion. In the contract, the owner should tailor the time required for a private entity to operate the facility to its ability to train current staff or retain additional staff. Often, the procurement documents will contain two or more time frames of operation from which the owner can choose at the time of procurement. Another approach includes a fee for extending the time of operations during the time of initial operation. Lastly, many in our industry believe that, in certain cases, private operations firms offer the knowledge, experience, and staffing capabilities that a municipality may never be able to attain. An operations element would then be part of the municipalities' overall approach to the operation of more complex municipal functions.

Another reason an owner might consider including an operations element would be to encourage the construction contractor to innovate, but with quality material and equipment. By making the contractor also responsible for a term of operations and having that cost become part of the bid evaluation, that contractor is discouraged from an approach with a low construction cost without concern for operations and maintenance (O&M) costs. This may be a significant benefit to the owner through obtaining a best combination of low initial cost with reliable, low-cost O&M. When using this approach, the time frame for operations by the private entity may be relatively short.

4.3.2 Advantages and Disadvantages

The following are the advantages of DBO delivery of a project:

- Allows owner with inadequate funds for construction to complete a necessary project in a timely fashion
- Allows owner with inadequate operations staff to begin operation of a WRRF in a timely fashion
- Assigns risks associated with ownership or operation to someone other than the municipal entity
- Encourages the construction contractor to pay more attention to quality when the contractor will be responsible for the cost of ownership or operation, especially with long-term operating contracts
- Encourages innovation by the bidder through the input of widely experienced constructors and operators during the bidding process
- Encourages cost-effective approaches through assessment of construction and operations costs during bidding, resulting in lower user charges
- Assigns the responsibility for design, construction, ownership, or operation, or any combination of these items, to a private entity

Disadvantages of the DBO approach for a constructed project include the following:

- Lack of familiarity with the selected approach by owners, contractors, and the public, especially in the United States

- Requires well-defined duties, responsibilities, and roles, which can complicate contract terms
- Loss of control by the municipality of any of the aspects assigned to the firm with whom the owner contracts
- Municipality may miss training for the operation of the facility, and there is potential for increased cost at the completion of the initial operation period
- User costs may be higher and are not as directly controlled by the municipal entity
- Loss of direct control over operations for which the municipal entity is still responsible through its discharge permit
- Quality may be compromised by low bid requirement
- Relies on the solvency and continued service of a private company
- Poorly defined duties and responsibilities can lead to significant conflicts
- Philosophical differences can arise regarding normal and emergency operations, maintenance, and public relations
- Bidding process and selection of private firm can be complex and time-consuming
- More applicable to larger, major projects

5.0   CONSTRUCTION MANAGER AT RISK DELIVERY

The use of construction management on municipal WRRF projects has been common for decades. Unfortunately, in the construction industry, the term "construction management" has widely varying definitions. An incredible range of roles, responsibilities, and authority of the parties involved in a project have been described using the single term of construction management. In addition, variations of what many would refer to as construction management have been developed to define the roles and responsibilities of owners, engineers, construction contractors, and construction management consultants.

It is beyond the scope of this manual to explore the wide variety of construction management forms and approaches. However, one particular approach has found recent favor in the wastewater industry. It is frequently referred to as CMAR. Other common terms in this approach include construction manager for general contractor or construction manager for builder. Because a number of more recent municipal WRRF projects have used this alternative delivery method, it will be described here.

One of the major intents of the CMAR approach is to combine the best of the design–bid–build and design–build processes. In this delivery method, the owner, either with internal staff or through a consulting engineer, develops an independent design. Somewhere before initiating the design or early in the design effort, the owner retains the construction manager. The construction manager is hired based on qualifications, experience, and other factors that the owner deems most important. Typically, at the time the construction manager is hired, the project is not yet defined well enough to provide a firm, fixed price for construction.

5.1   Sequencing of Tasks

The reason for retaining the construction manager early in the design process is to use the manager's design knowledge and skills to improve the schedule, constructability, and quality of the project while retaining construction cost control. Then, at some defined point in the design process (anywhere from 30% to 90% design complete) the owner and construction manager, with input and assistance from the design engineer, negotiate a firm, fixed fee for project construction. One contractual approach for this is to use a two-part contract with the construction manager. The first part defines the construction manager's roles, responsibilities, authority, and fees during the design process. Fees for the first part can be based on a defined scope of services and duration—in other words, paid consulting services. The second part, for which the fee is not defined until much later, is for actual project construction. The construction manager may initially submit their fees for management services during the construction period as a percentage of the ultimate GMP. The second part may or may not require the construction manager to complete the design or to provide additional design drawings, similar to the design–build approach.

A major concern public owners have with this approach is the appearance of receiving a high construction cost with no direct alternatives or bids to compare because that portion of the project is not competitively bid, as it is in the design–bid–build process. One way to mitigate this is for the construction manager to either "open the books" on final construction fee preparation or to competitively bid on certain major subcontracts. The advantage to the owner is a better sense of cost control. Loss of control over selection of subcontractors can affect a construction manager's ability to provide quality and save time. As a last resort, it would be possible at this time to bid the project as a traditionally delivered design–bid–build project if the construction price is unfavorable. Section 5.2 summarizes the typical responsibilities of each party in the CMAR approach.

5.2   Project Team and Primary Functional Relationships

The following are typical responsibilities for each party in the CMAR delivery method:

- The owner's responsibilities in the CMAR delivery process are essentially the same as in traditional project delivery systems. The owner is responsible for:
  - Determining the goals and requirements for the project
  - Acquiring a usable site for the project
  - Financing the project
  - Selecting the engineer and managing the bidding or negotiation to select the construction manager
  - Directing the engineer and the construction manager
- The design engineer is responsible for:
  - Predesign
  - Assisting the owner in determining project goals and requirements
  - Developing the project's design
  - Coordinating the design consultants
  - Processing entitlements related to design responsibilities, such as planning approvals and zoning variances
  - Ensuring regulatory and code compliance
  - Preparing scope-of-work documents for the construction manager selection
  - Assisting in the bidding or negotiation process to select the construction manager
  - Working with the construction manager during constructability reviews
  - Preparing construction documents
  - Reviewing and approving shop drawings
  - Administering the contract for construction
- The construction manager is responsible for tasks of management as well as construction and is responsible for:
  - Preparing estimating and scheduling data during the design phase
  - Providing value engineering and constructability reviews during the design phase
  - Managing bidding or negotiation for the work of construction
  - Contracting for the cost of construction

- The CMAR is responsible for the construction, including the hiring of prime or trade contractors who carry out the actual construction under contract to the CMAR. Construction responsibilities include:
  - Committing to the cost of construction, which typically includes bids of the actual cost of construction from prime or trade contractors
  - Gaining entitlements related to construction, such as building and encroachment permits
  - Preparing shop drawings and other documents necessary to accomplish the work
  - Managing the construction process, including establishing all contracts with prime or trade contractors
  - Establishing and maintaining the construction schedule
  - Providing the methods and means of construction
  - Ensuring job-site safety
  - Fulfilling the requirements of the construction documents
  - Guaranteeing the quality of construction
  - Correcting deficiencies covered by the guarantee

5.3   Uses

Many owners face material and equipment quality problems associated with the low-bid process in the conventional design–bid–build delivery method. It is one of the reasons alternative delivery methods have received attention and use. One major reason an owner would consider use of the CMAR approach is to achieve much better final project quality through increased owner control of material and equipment selection.

By retaining the independence of the design engineer while eliminating the hard bid requirements of the design–bid–build process, the owner appears to get the best of both worlds. The owner is still able to exert control and influence over the design through control over the design engineer, which is a major advantage of this approach compared to the design–build approach. The owner can better control project quality by controlling construction through the contract with the construction manager. This is believed to be a major advantage of CMAR over the design–bid–build delivery method. And, hopefully, the owner expects to enjoy adequate cost control through the various cost checks that can be incorporated into the construction manager's contract and the independent checks provided through the design consultant.

5.4   Procurement

The first step in a CMAR procurement is typically to advertise the preconstruction services portion of the project. This is a relatively small level of preparation because, at this time, the project does not need to be advanced in its design progression. A key challenge is planning and education internally on how the CMAR selection process should occur. As mentioned previously, the price components submitted are frequently both a dollar amount for pre-construction services and a percentage for the construction-phase services. Agencies must establish ahead of time how these values will be compared to prevent any opportunity for protests from respondents not selected and the need to restart procurement.

5.5   Advantages

The advantages of the CMAR alternative project delivery method include the following:

- The construction manager provides valuable input during the design on issues such as schedule, budget, quality, and constructability.
- The opportunity exists for the construction manager to begin construction earlier, shortening overall project delivery time.
- There are engineering cost savings by not having to take the design to 100% completion.
- The construction manager can reduce the owner's management burden during construction, serving as both management assistant and construction contractor.
- Cooperation during the design phase significantly reduces the potential for disputes and litigation against the design engineer.
- Negotiating the construction fee enhances the owner's ability to control the quality of the final constructed project.
- The independent design engineer still provides the owner with a system of checks and balances.
- Because the designer and contractor are retained under separate contracts, lines of responsibility, authority, and liability are similar to design–bid–build and, therefore, more familiar and easier to define.

5.6   Disadvantages

Some of the disadvantages of the CMAR project delivery approach include the following:

- The process is unfamiliar to most municipalities and contractors.
- Adversarial relationships could remain between the engineer and construction manager.

- Disputes can arise regarding the completeness or accuracy of the documents provided by the designer.
- There may be a perception of not achieving the desired quality.
- The potential for change orders still exists.
- The role of the construction manager can appear contradictory, that is, both advising the owner on the project and constructing it.
- There is additional cost during the design effort for the input of the construction manager.
- Because the process is still mostly sequential, the potential for time savings compared to a design–build project is reduced.
- The owner has less control over the actual construction contractors or major subcontractors because there is no direct contractual relationship with them.
- Because the project is not competitively bid, there may be the appearance of no cost control over the construction cost, especially to the public.
- The method is primarily applicable to larger, diverse projects.

5.7   Risks

This section will provide a brief overview of key risk considerations in the CMAR delivery method.

5.7.1   Role of the Designer

The main benefit of designer involvement is to shape the final design documents so that both parties can agree on an accurate GMP for the construction. However, because the designer is not working to advance design and does not benefit from feedback from contract packages before proceeding into construction, there may be significant rework once site conditions are known.

5.7.3   Role of the Contractor

The contractor's supporting role is helpful in advising the designer on constructability and other considerations that make for a more robust design. However, setting a GMP may create an unreasonable risk allocation that the contractor must mitigate.

5.7.3   Expectations for Collaboration

Collaboration is key to alleviating these concerns in a CMAR project. Through early communication and working together, cost control from the start of the project can be achieved. The public owner has great visibility into what the project will really cost, but this could result in risk to the project: cancellation of the project mid-stream once it becomes evident that the project does not fit within the allocated funding.

5.8   Limitations

CMAR is not well suited for all projects. In particular, CMAR is not a good fit for smaller projects. The CMAR structure can also result in particular aspects of the contract structure that expose the client to liability. Unlike in a design–build project, the owner is fundamentally responsible in CMAR for the interaction between designer and contractor, and any associated gaps that create issues in that relationship. One other limitation is that the CMAR entity may not be on board in time to be part of designer selection. In these cases, the relationship does not benefit from designer and contractor arrangements and the advantages of working together on past projects like in a design–build project.

6.0   PUBLIC–PRIVATE PARTNERSHIPS DELIVERY

6.1   Uses

The public–private partnership is neither a new concept nor a radical delivery method. In the United States, the use of public–private partnerships dates back several hundred years, although in the 1900s so-called "traditional" delivery methods became more common. While public–private partnership contracting methods vary, there are some common advantages and disadvantages to consider with each project delivery type.

Public–private partnerships can be further split into types based on the degree of private participation. Some experts would consider many of the delivery methods discussed earlier in this section as types of public–private partnerships. The most severe forms of public–private partnerships are design–build–finance–operate–maintain contracts (DBFOMs) or full-system privatizations. A major distinction is whether a project's payment mechanism is based on user fees or availability payments.

6.2   Due Diligence

Navigating the complex waters of a public–private partnership transaction requires the right team in the owner's corner. The roles of the technical advisor, legal advisor, insurance advisor, and transactional advisor are diverse and must all be appropriately staffed with experienced team members. An important consideration is the role of chief coordinator; depending on how this role is staffed, the chief coordinator may result in many organizational advantages to the client. It is also important to consider the experience that the advisors have in working together and in the area of infrastructure. Knowledge of industry players, media groups, the legislative process, and other details can be helpful in ensuring that a proposed project turns into a successful solicitation.

6.3   Financing

More comprehensive public–private partnerships are characterized by contracts that integrate design, construction, financing, and/or O&M elements through a single contract, often over a lengthy concession term. A special-purpose vehicle (SPV) is often established as a legal entity solely for the project. This structure limits the liability of investors and makes the project less susceptible to the liabilities of the firms making up the SPV. Assets and liabilities of the SPV are only those related to the project. The SPV receives equity contributions from the firms that are equity members and pays dividends to the investors in return.

6.4   Compliance

Compliance considerations are key in public–private partnerships, and the long-term debt as well as potential tax implications create unique scenarios. A thorough analysis of the governing structure at each utility with comprehensive advice across technical, financial, legal, and commercial teams is key to a successful public–private partnership that does not create future compliance issues. Another important consideration is compliance with the concessionaire contract in the future. Most owners are not accustomed to the level of oversight that is appropriate for alternative delivery methods with large private sector involvement like DBFOMs. Understanding what aspects of quality oversight are appropriate to assign to the project developer is key.

6.5   Debt-to-Equity Ratio

A public–private partnership transaction with private financing incorporated will require extensive financial modeling and due diligence. In any transaction with financing, considering the debt-to-equity ratio is key to understanding the financial health of a transaction. The financing entity and the public sector project sponsor will negotiate the terms. While entities providing equity seek to minimize the investment, public sector owners must be careful to ensure that sufficient equity investment is provided to prevent rapid turnover in the asset's ownership over time.

6.6   Advantages and Disadvantages

There are several advantages to the public–private partnership concept. Public–private partnerships provide a complete life-cycle approach to infrastructure project development. By incorporating long-term O&M, a public–private partnership developer is incentivized to construct a truly efficient and future-thinking facility. And particularly for specialized technologies for which a public owner may not have operating experience, the notion of contracting O&M to the same entity responsible for design and construction is appealing.

There are some disadvantages to the public–private partnership delivery method. Many successful public–private partnership projects have been for larger projects where the transaction costs are small relative to the total project cost. Often, the transaction costs associated with public–private partnership projects are not proportional to project cost, yielding economies for larger transactions. While attempts have been made to aggregate projects into packages to reduce transaction costs and make the public–private partnership model more viable for smaller system projects (West Coast Infrastructure Exchange, 2012), this delivery method is not commonly adopted.

7.0   PROCUREMENT AND DELIVERY

Chapter 9 of this manual will discuss the principles for successful project completion, and can be referenced for projects using all the delivery methods discussed in this chapter.

7.1   Risk Management

In this chapter, key risks are identified alongside each delivery method. Alternative delivery contracting presents benefits to owners in allocating risks on projects—in many cases, contract language can be structured such that certain risks are borne by the private sector rather than the owner. The key to owners using this information successfully is developing a keen understanding of risk allocation and considerations important to this commercially sensitive part of the process.

All risk transfer comes at a price to the public sector owner. Transferring too much risk, or risk that the private sector cannot reasonably expect to control, can result in wild cost increases. However, transferring to the private sector only the appropriate risks can result in long-term life cycle cost advantages that can mean what initially presents as a "higher" price for a project upfront may result in significant efficiencies and savings.

7.2 Legal Considerations

As has been noted throughout this chapter, each project delivery method has specific requirements for the proper preparation of procurement, bidding, or other project documents. In the United States, most of the regulations affecting a municipality's selection of a project delivery method are based on state law. In many cases, local law can modify or supplement the state requirements.

It is imperative that the owner ascertains that the project delivery method selected is legal within that state. For example, the design–build approach is currently not legal in many states for municipal projects. The same laws may not constrain sanitary districts or authorities. Also, many states have exceptions for selecting approaches such as design–build under emergency or other similar conditions. Because state laws and regulations are constantly changing, owners may wish to frequently assess what is allowed in their states.

Regardless of the delivery method selected, it is important to ensure that the various possible contracts and project documents adequately define the roles, duties, responsibilities, and authority of each of the parties throughout the duration of the contract. Some key issues to consider include the following:

- Owner's right to approve or change the design
- Owner's right to approve or change schedules
- Audit rights of any party
- Authority over cost final construction cost
- Mechanisms for conflict resolution
- Duty of care
- Warranties
- Assignment of liability
- Limitation of liability
- Compliance with codes and law
- Compliance with terms of discharge permit
- Responsibility for construction site safety
- Responsibility for safety of operations staff
- Management and control of subcontractors and suppliers
- Bonding requirements
- Insurance requirements, provisions, limits, and naming of insured

- Indemnity provisions
- Right to suspend, terminate, or replace
- Penalties, liquidated damages, actual damages, and bonuses
- Responsibility for environmental hazards and damages
- Responsibility for permit violations, fines, and orders

It is also important to consider how any of these items may change as elements of the design, construction, and operation are initiated or completed.

As has been mentioned previously, there are numerous organizations with manuals and standard documents that deal with the above issues. Legal counsel should be sought whenever questions arise regarding properly addressing any contract issue such as those listed above.

8.0   SUMMARY

A critical element of the municipal owner's completion of a WRRF project is the selection of the method under which the project will be designed and constructed. Aside from the commonly used design–bid–build method, other alternative project delivery systems are available that may be of great benefit depending on individual project conditions. Careful consideration of all available methods will help ensure that the municipality receives the best value for its expenditure.

9.0   REFERENCES

United States v. Spearin, 248 U.S. 132, 39 S. Ct. 59, 63 L. Ed. 166 (1918). https://www.loc.gov/item/usrep248132/

West Coast Infrastructure Exchange. (2012). *West Coast infrastructure exchange final report.* CH2MHILL. https://digital.osl.state.or.us/islandora/object/osl:16626

10.0   SUGGESTED READINGS

American Institute of Architects. (1996). *Handbook on project delivery.* American Institute of Architects, California Council.

Association for the Improvement of American Infrastructure. (n.d.) *Guide to successful P3 evaluation and delivery.* https://aiai-infra.info/resources/guide-to-successful-p3-evaluation-and-delivery/

Bender, W. J. (2007). *Defining and allocating "design responsibility" in complex projects*. Skellenger Bender PS. https://www.skellengerbender.com/wp-content/uploads/2017/10/2007-Defining-and-Allocating-Design-Responsibility-in-Complex-Projects.pdf

Construction Management Association of America. (2002). *Construction management standards of practice*. Construction Management Association of America.

Design–Build Institute of America. (1996). *An introduction to design–build* (DBIA Manual of Practice No. 101). Design–Build Institute of America.

Design–Build Institute of America. (2013). *Design–build done right: Best design–build practices*. https://dbia.org/wp-content/uploads/2018/05/Best-Practices-Universally-Applicable.pdf

Frederickson, K. (1998). Design guidelines for design–build projects. *Journal of Management in Engineering, 14*(1), 77–80.

Raftelis, G. A. (2014). *Water and wastewater finance and pricing: The changing landscape* (4th ed.). CRC Press.

Water Research Foundation. (2009). *Improving water utility capital efficiency*. Water Research Foundation.

6

Value Engineering and Constructability Reviews for Facilities

Tina Wolff, PE & Kameryn Wright, PE

1.0   INTRODUCTION

Capital improvement projects for facilities start with broad concepts and narrow to highly specific scopes of work. This is a necessity, of course, as contractors cannot price and build vaguely described projects. But in narrowing the lens, aspects of a project critical for a long, productive life can be less than fully integrated into the work. Examples of such aspects include operability of equipment or valves, compatibility of new technology with old, technology advancements, and buried costs for maintenance. These are the types of issues that can be identified by value engineering and constructability reviews. This chapter was placed here specifically to raise awareness among owners and engineers of the value of third-party reviews so that they can plan reviews during the project process.

One common owner question is: why should owners perform value engineering and/or constructability reviews—isn't that why they hire a design engineer? Everyone invested in the project, including the designer's team and the owner's team, will become so ingrained in the details of the work that it becomes impossible to be objective. The project reaches the point where the teams can't see the forest for the trees, to use an old saying. Value engineering and constructability reviews bring in fresh eyes and new ideas, not with the intent of tearing down the good work of the project team, but to identify all the bits and pieces that can make a good idea great. The main purpose of value engineering and constructability reviews is to maximize the functionality, performance, quality, and operability of a system for each dollar invested. It should be noted that value engineering and constructability reviews are not lifelines for projects for which the scope has outpaced the budget.

Value engineering is a structured process that uses a facilitator to analyze a project from the points of view of industry professionals across multiple disciplines to maximize benefits while lowering construction costs, operational costs, and risks. A constructability review is a less formal process and uses construction professionals to assess the viability of proposed construction means and methods, and focuses on the quality of the facility being delivered while lowering construction costs and risks. Typically, system function, performance, and operability are not included in constructability reviews, and the overall project intent remains unchanged.

2.0   DEFINING VALUE ENGINEERING

Value engineering is a systematic approach for identifying a project's functional and value objectives with the goal of optimizing design, construction, and future operations versus cost. Value engineering studies are conducted by a multidisciplinary team of design, operation, and construction professionals who evaluate the work done by the project team.

Value engineering originated in the 1940s and was adopted for water and wastewater projects in the mid-1970s. Value engineering has since grown from a study during the design phase to an approach for delivering projects with value intents that recognize the importance of value versus cost cutting. The success of value engineering is in communicating the owner's value objectives and attaining those goals.

In the execution of federal work, value engineering is a standard part of the acquisition process. It is governed by Federal Acquisition Regulations (FAR) Parts 48 and 52.248-1 (FAR Part 48, 2020, FAR Part 52.248-1, 2020) and encourages the contractor to voluntarily identify opportunities for savings while preserving the intent of the design. The contractor may share in the savings, thus having an incentive for undertaking the work.

2.1   Goals of Value Engineering

The goal of value engineering is to maximize the value to the owner by balancing benefits against cost and to find an optimal point where the most value is gained for the dollars spent. Value engineering does not seek to achieve the lowest cost of construction by sacrificing quality, efficiency, or operability. For example, value engineering may not recommend purchasing an alternate, less expensive pump with poorer operating efficiency and higher average maintenance cost compared to other viable alternatives. Although the alternate has a lower initial cost, its cost over the lifetime of the pump may significantly outpace the other alternatives. Not all benefits can be evaluated by cost alone, though. Take, for example, two sludge dewatering systems that provide equivalent dewatering capability at roughly the same cost over their lifetime. Although these two systems have the same relative cost, one system may create an undesirable environment during operation that lowers the value of that system for the owner/operator when compared to the other dewatering system. A further illustration of this point is comparing styles of aeration blowers with equivalent cost and performance, but producing varying environmental conditions during operation as a result of the number of decibels produced by each blower type.

Conversely, value engineering does not seek to maximize the benefit disproportionate to cost. A good example of this is installing electrically actuated gates in an open channel. If the sole function is channel isolation for infrequent maintenance, this may have been achieved just as effectively and more economically with a manual gate or stop plate.

2.2   Relationship to Project Budget

As previously mentioned, value engineering is not a lifeline that can fix a flawed budget. Rather, value engineering will demonstrate how to get the most done for the available budget. In a situation where the project is significantly over budget (greater than 15%), the expectation should be that scope must be cut. Typical value engineering studies result in 15:1 to 25:1 cost savings, or approximately 4% to 6% of the construction cost. Results will vary with project size and complexity.

As can be seen in Figure 6.1, value engineering is typically most effective when performed in the early stages of design. Renovation projects may differ from this trend and can often be the most difficult type of construction. Renovation projects often require facility operations to be maintained during the disruptions of construction, with a greater potential for rework, lack of clarity regarding existing conditions, and the problematic merging of old and new infrastructure. For these reasons, renovation projects may have significant value engineering opportunities later in the project process as more unknowns become uncovered.

3.0   PLANNING FOR VALUE ENGINEERING

Value engineering studies benefit project quality at all stages of development. Traditionally, value engineering studies are undertaken at one or more of the following project stages:

- At the conclusion of concept or facilities planning (tests the basic program and validates decision-making)

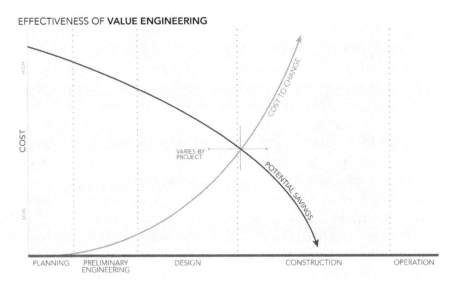

FIGURE 6.1   Relationship between cost and benefit to project changes.

- At 30% design completion (design, layout, process, and overall concepts)
- At 60% design completion (details of the applications of the work, reviews for maintenance of facility operations, and project delivery provisions)
- At 90% or 100% design completion (thorough review of the plans and specifications for constructability, biddability, operability, and maintainability)

Value engineering studies performed during the early stages of a project tend to address broad project concerns related to the selection of the right design concepts and elements. The identification of effective solutions during conceptual planning tends to provide the largest savings in life cycle costs. Value engineering is a proactive effort to improve a project. This proactive strategy also saves on additional design fees.

Later studies help to fine-tune the design and design operability. The later reviews have the greatest effect on future operations and the quality features of the project. Value engineering studies performed in the latter stages of a project tend to focus on getting the best value from the project elements that have been selected.

The length of time that the value engineering team works together in a workshop varies with the size, nature, schedule, cost, and complexity of each project.

TABLE 6.1 Value engineering study guidelines.

Project Construction Cost/ Complexity	Value Engineering Application
Less than $5 million and uncomplicated	Value engineering studies offer the most benefit during the conceptual stage and with constructability or operability assessments before construction.
$5 to $15 million and uncomplicated	One value engineering study when design is approximately 20% to 30% complete typically offers the most benefit.
Large (more than $40 million) or highly complex	Two value engineering studies typically provide the most benefit: one at the concept phase (10% to 20% design complete) and the second when design is approximately 65% to 75% complete, with a constructability/ operability study before construction.

Note. For large projects, workshops are often held to discuss individual components as they are designed.

3.1   Facility Planning/Conceptual Design Phase

In this phase, value engineering involves validating the solution concept. The project team works to identify the owner's requirements and the concepts that best achieve them. Benefits of early application include:

- Consensus for the process and design
- More realistic budgets
- Early evaluation of construction sequencing to maintain facility operations during construction
- Early operations and maintenance input

Designs are then developed with a cost-budgeted approach for capital and operations costs, and designs are measured against these goals. Value engineering during the conceptual stage may be best suited for projects that are significantly complex, propose novel solutions, or have long gaps of time between proposing and implementing the project. A prime example would be a municipality having a long implementation schedule for its combined sewer overflow long-term control plan. Depending on circumstances, some municipalities can expect 20 years for full implementation, by which time significant advances in technology can potentially provide more valuable alternatives.

3.2   Preliminary Design Development Phase

Value engineering may be applied during the preliminary design development phase (approximately 10% to 30% of the design effort) where changes resulting from value engineering studies can be implemented without significant effects to the project schedule or design budget. The potential for value engineering–related cost savings is substantial during this period. Among the benefits are overall optimization of the design, confirmation of the owner's goals and objectives, and validation of the design approach.

3.3   Final Design Phase

Value engineering studies undertaken in the final design phase (60% completion or more) should strongly focus on operational improvements, finishes, and quality standards for equipment and materials. Major changes to the project intent or scope that result from value engineering studies should be expected to affect both the project schedule and the design budget. As a result, the savings potential must be carefully weighed against the cost of redesign and schedule delays.

More commonly, work at this phase focuses on design implementation. For example, an allowable level of disruption to facility operations during construction can be established and suitable construction methodology determined. The constraints associated with maintaining facility operations can have a significant effect on the cost of construction. This provides another opportunity for savings, especially on projects with limited budgets.

3.4 Preconstruction Phase

Value engineering studies performed during the preconstruction phase should focus on optimizing the contract documents for biddability, minimization of contractor risk (real or perceived), expectations for maintaining facility operations during construction, and clarification of work restrictions such as weekend or night work. Value engineering performed during this phase should aim to reduce contract change orders and unnecessary claims. This is especially important for rehabilitation projects where the risks from unforeseen conditions are higher.

Another potential value engineering opportunity may come from equipment or material market volatility. A prime example is the effect Hurricane Harvey had on the availability and price of polyvinyl chloride pipe in late 2017 because of the major disruption the hurricane caused to manufacturing facilities located near the south coast of Texas (Sherman, 2017). In this example, allowable pipe materials could be expanded in the contract documents and/or alternative bids developed to overcome material volatility.

Major changes resulting from value engineering studies during the preconstruction phase should be expected to affect both the project schedule and the design budget. The savings potential from these changes must be carefully weighed against the cost of redesign and schedule delays.

3.5 Construction Phase

Value engineering studies undertaken in the construction phase are called Value Engineering Change Proposals (VECPs). Major changes resulting during this phase will likely affect the project schedule, the construction budget, and the design budget. Generally, the savings realized by implementing VECPs in the construction budget are several times larger than the cost to engineer the solution, making time the bigger factor. VECPs commonly share the adopted savings between the owner and constructor, after deducting expenses related to engineering development and review of the proposed change.

4.0 VALUE ENGINEERING AS A DISCIPLINE

Value engineering is more than an industry term for optimizing major capital projects and programs. It is a specialized discipline and can be considered similar to electrical engineering or structural engineering. Value engineering professionals are certified with SAVE International, which advances, promotes, and accredits individuals in value methodology. Founded in 1959 as the Society of American Value Engineers (SAVE), SAVE International recognizes competence in the practice of value methodology.

Fundamental to the concept of value is using a single term—the value index—to express both the benefit and the cost elements of a project. The value index aims to quantify a project's value by determining the relationship of worth versus project cost. This relationship is expressed as follows:

$$Value\ Index = \frac{Function + Performance + Quality + Operation + Maintenance}{Cost + Risk}$$

A well-executed value engineering process will apply this test to the project's concept or design, depending on the phase, and identify strategies to maximize the value index. The numeric ratings for each factor are set within each project. The key is consistently reflecting the owner's priorities in application throughout the project. For example, an owner who intends to sell a facility within 5 years of its completion may place less value on long-term maintenance characteristics than an owner who intends to keep a facility for decades. Likewise, severity of fines for exceeding effluent limits will influence the robustness of the design. Owner requirements directly affect the relative value of aesthetics, reliability, sustainable development, maintainability, operability, construction duration, and other characteristics. Improving the quality of elements related to these characteristics generally increases cost. The goal of value engineering is to achieve a ratio of quality to cost/risk that is acceptable to, and in the best interest of, the owner.

5.0   VALUE ENGINEERING PROCESS

The following sections outline the basic process for value engineering to assist owners who are new to the value engineering process with planning. In addition to the information provided, owners are encouraged to use their own network of peers and professionals to inform their process.

5.1   Budgeting for Value Engineering

Owners should expect to incur a cost for value engineering and should plan for this in the project schedule. The project schedule should not only budget the time to conduct value engineering but also the time it will take to implement changes resulting from value engineering. Costs for value engineering depend on the number of people participating, costs for participation, and the number of value engineering workshops completed. Table 6.2 identifies common cost elements for consideration. Conceptually, costs for value engineering services can be 0.4% of construction costs for larger projects and up to 0.8% of construction costs for smaller projects.

TABLE 6.2   Value engineering basis for budgeting.

Category	Hours	Cost / Hour	Travel Expenses	Accommodations / Meals
Value engineering facilitator	120 hours including 5 days/ 40 hours on site	$200–$400	Varies	$200–$300/day for each day on site
Value engineering support staff	120 hours including 5 days/ 40 hours on site	$100–$200	Varies	$200–$300/day for each day on site
Subject matter expert (remote)	5 days/40 hours	$200–$400	Varies	$200–$300/day
Subject matter expert (local)	5 days/40 hours	$200–$400	N/A	$40/day
Subject matter expert (owner staff)	5 days/40 hours	$25–$200	N/A	$40/day

Table 6.2 provides a starting point for developing a budget for value engineering. Hourly labor costs and expenses vary significantly across the country. Owners and engineers are generally expected to know the costs in their regions. Subject matter experts who work for the owner can and should participate. Often, the costs for these team members are neglected, as they are not invoiced to the owner.

5.2   Constructing a Diverse Team

To be effective, the value engineering team must have a diverse background that cover the key disciplines (process, civil, structural, electrical, instrumentation and control, mechanical, construction, operations, etc.) involved in the project. Value engineering teams benefit from members with a range of expertise in the project's key issues. Positive attitude, technical knowledge, education, certification, and professional experience are also desirable qualities for each member of the team.

5.3   Preparatory Review

The facilitator will require time to become familiar with the project and plan the approach to the value engineering. The owner should expect to make all current and past documents available to the facilitator. Copies of key documents will need to be made for all participants. The size of planning documents, costs of color printing, and other factors entail planning for these costs as well.

5.4   On-Site Workshop

The on-site workshop typically begins with a presentation of the project by the project team to the value engineering team. The value engineering team may ask questions to get a better understanding of the owner's drivers and priorities, site constraints, schedule constraints, and other considerations. After the presentation, the value engineering team is generally sequestered to begin their work. It is common to have the value engineering team hosted off-site, such as in a conference room at a hotel. The value engineering team will need access to common office supplies, internet, and other resources as identified by the facilitator. Because of the amount of work to be done in a short period of time, the value engineering team often works long hours. Consideration of meals and other comforts needs to be included. The workshop concludes with a presentation of value engineering concepts to the design team. The degree to which the concepts are developed will vary with available time and resources. Generally, the concept will include an order-of-magnitude estimate of cost to implement and cost savings. All concepts should be expected to improve the value index over the score for the base concept.

5.5   Leveraging Results

Once the value engineering concepts are turned over to the owner, the owner and engineer should expect to work to fully vet these concepts, including performing basic calculations, developing or confirming cost estimates, and vetting the perception of risk reduction. Ultimately, it is the owner's decision regarding which, if any, recommendations to incorporate into the design. Both the owner and engineer should be aware of the natural tendency to become defensive in light of differing ideas. Both are reminded that the goal is to improve on good ideas, not to be critical of the work performed.

5.6   Managing Value Engineering Costs

As noted previously, value engineering can be costly to implement, and yet the benefit to projects of all sizes can be tremendous. For owners with construction budgets under $15 million or those wishing to minimize costs, the following considerations are offered:

- Invest in the facilitator; this is the person who drives the success of the exercise.
- Bring in local subject matter experts to lower travel expenses.

- Consider retirees as subject matter experts; they bring decades of experience to the table and, working as individual consultants, may save the costs of company overhead.
- Use space within the facility or at another owner property to lower overall costs.

6.0 CONSTRUCTABILITY REVIEWS

Where value engineering considers the overall benefit provided by the project cost, constructability reviews focus on the practicalities and risks of constructing a facility. A constructability review is a tool used to deliver a set of improvements for a fair industry price. Generally, consideration for operability and maintainability is limited to the selection of materials, products, and methods that enable a system to achieve a full-service life. It should be noted that project delivery methods for which the contractor is selected during the design stages, such as design–build and public–private partnerships, can inherently include contractor input regarding constructability concerns and/or risk. Project delivery methods that do not have contractor involvement during the design process, such as design–bid–build, will have to explicitly introduce a constructability review.

6.1 Goals of Constructability Reviews

In the biggest sense, the goal of a constructability review is to increase confidence that the project can be constructed under the established budget. To get to this point, constructability reviews consider many points, including the following:

- Safety: For contractors, safety is paramount. Each contractor will incorporate the work needed to create a safe work environment into the price. This includes normal activities, such as cutting back excavations at a stable slope, to extraordinary activities, such as purchasing custom equipment.
- Nonproduction schedule: The cost of construction can be affected by the scheduled construction productivity, including environmental constraints (such as a limit on tree clearing), regulatory constraints (such as seasonal disinfection requirements), and financial constraints (such as loan closing dates or requirements to spend funds).
- Sequence constraints: The lowest-cost projects will be those that allow a contractor to aggressively execute the work. Projects with sequence

constraints slow contractors down, drawing out work and costs. Notably, general condition costs can be significantly higher because of the longer schedule.

- Specialty equipment requirements: Deep excavation and high installations are examples of two conditions that can require specialty equipment to execute the work. Some contractors may have some of this equipment as part of their fleet. Others will rent what is needed. In some situations, contractors may have to innovate and create a system that isn't currently available. This can drive costs beyond the simple cost of the work.

- Identification of unknowns: Engineers and owners do their best to identify existing conditions. But frequently, contractors will ask questions critical to pricing that are unknowable. How much groundwater will need to be pumped? How thick is the sludge in the digester, and can it be pumped? Without incorporating some basis for pricing, the contractor is put in a position of having to cover costs for an unknown.

- Availability of key equipment and materials: Many critical wastewater components have lead times approaching 26 weeks. Considerations such as the American Iron and Steel provision and a busy construction environment can further stretch out schedules and costs.

- Seasonal impacts: Around the country, the different seasons can have real effects on construction costs and productivity. In winter, work can be either shut down or require additional steps to protect workers and products from inclement weather. Wet seasons can affect daily operations, but from a constructability review standpoint, the effects typically considered are longer term, such as a high groundwater table.

- Complexity of structures: Every bend in a pipe or wing of a building costs money. Reviews will often look for ways to simplify a design to lower costs.

- Industry disruption: More difficult to take into consideration are disruptions to the industry. Hurricanes, wildfires, flooding, virus outbreaks. All these natural events have disrupted the construction industry supply chain in recent years. Constructability reviews can help identify where projects are at risk for delays or increased costs because of a disruption in the overall system.

- Competition: Owners know that contractors are in competition with each other. It is also true that one owner's project is in competition with a neighbor's project. Competition on both fronts can invisibly affect a project's costs. It is important to understand where your project fits in and, if possible, adjust timing for the best pricing.

6.2   Relationship to Project Budget

As with value engineering, constructability reviews will not solve problems of an underfunded budget. Constructability reviews will help an owner and engineer understand where project costs appear out of line with initial concepts, and where contractors see risk. It is in managing these two areas that the bid price may be reduced. The amount of cost that can be saved varies with the complexity of a project. For simple, straightforward projects, constructability reviews may identify 2% to 3% in savings. For complicated projects, the potential can reach 10% to 15%. Where budgets are underfunded by 10% or more, owners should expect to add funds or cut scope.

6.3   Applicability Across Project Sizes

The value of constructability reviews does not vary by size. Across the board, constructability reviews help owners understand how contractors perceive the design document, where contractors identify risks, and which project elements can be modified to lower costs.

6.4   Safety

Construction can be a dangerous field, and the long-term operation of a facility has its own set of safety risks. From the construction standpoint, the constructability review can identify places where safety risks push pricing outside standard ranges and offer alternatives to improve construction safety and, potentially, lower costs. For example, consider a deep sewer excavation. On a pure diameter or linear foot basis, open cut may appear less expensive than a trenchless technology. But when you factor in the extraordinary measures that could be required for trench safety, especially in poor soils, plus all of the additional costs for stone, dewatering, and other factors, the trenchless installation may become more financially viable and is certainly safer. There are likely additional measures required for trench safety whenever you are in an urban area for maintenance of traffic, signage, barricades, flaggers, and other requirements that further drive up costs.

6.5   Criticality of Intermediate Construction Stages

One area commonly undervalued when establishing a construction budget is the intermediate construction stage. Construction drawings give detailed information of the "before" and "after." The "during" is often left to the contractor, unless expressed as a constraint. As a result, costly activities may be overlooked in the development of a construction budget or engineer's estimate of construction. These activities generally fall under the category of means and methods, and a contractor's input can greatly help understand where hidden dollars lie. Additional information regarding this topic can be found in Chapter 9, Section 5.

Here is an example: a new tank is being added to an existing aeration basin. The effluent pipe is to be relocated from along the existing outside wall to along the new outside wall. As designed, the existing effluent pipe would have been required to be removed from service for the duration of time it took to construct the tank, which was several months. The original concept assumed the contractor could "work around" the pipe and did not include the cost of bypass pumping for the period.

6.6   Planning for Maintenance of Facility Operations

The requirement to maintain facility operations is typically put on the contractor with little information provided on what is required to maintain those operations. A good contractor will work with an owner to understand these operations, which may include the routine of facility operators such as when and where they need access and procedure changes, if any, in the event of heavy rain. For complex facility upgrades, closing and opening valves can have significant effects on both the owner and the contractor, adding unplanned costs and problems.

For example, for a WRRF upgrade, the engineer considered the hydraulic impacts at average and peak daily flow during construction, but failed to consider wet weather flow. The facility operational process for wet weather was followed, which caused a temporary weir to be overtopped, flooding the contractor's work area. The result was a week's delay in work as the tank was pumped back into the process along with costs associated with the delay, lost materials, and equipment damage.

Additional information regarding this topic can be found in Chapter 9, Section 5.

7.0   CONSTRUCTABILITY REVIEW PROCESS

Constructability reviews use construction professionals to assess the viability of potential or proposed construction means and methods. These reviews do not challenge the premise or generally offer a counterproposal to the solution, but look to implement the solution in the optimal manner. Constructability reviews are typically built into alternative delivery methods such as design–build and build–operate–transfer because the contractor is on board early in the project. This section focuses on constructability reviews for bidding, where the contractor reviewer is a noncontracted third party to the design.

7.1   Budgeting for Constructability Reviews

Constructability reviews can range from informal discussion of key areas to formal reviews of entire projects. It is equally common to have the contractor reviewer provide reviews at no cost as it is to pay for the time invested and expertise shared. In general, a constructability review that is limited to the time of a key senior team member and, perhaps, one other staff member is treated as a marketing and/or business development exercise. Constructability reviews that require significant time from estimators, project managers, and/or superintendents will likely fall outside that category. In these situations, contractor reviewers may be less willing to redirect staff off of active projects and/or active bids, even if paid for their time. Organizations are encouraged to think first about what they hope to gain from the constructability review and structure the request in as focused a manner as possible.

In the same vein as monetary budgeting, time must also be budgeted. Depending on the complexity of the project, the organizer should budget 2 to 3 weeks for the contractor reviewer to review the documents and develop a response. Scheduling a review meeting within that time frame nicely establishes a deadline and provides the opportunity to discuss areas of concern.

7.2   Targeted Expertise

It should be recognized that contractors, too, have areas of expertise. When performing constructability reviews, consider if a review is needed in areas generally considered specialty for wastewater construction. This includes electrical, instrumentation and controls, geotechnical piles, trenchless technologies, rock tunneling, and water/marine-based work, among others. Specialty work can be a substantial part of the overall contract value. In fields such as electrical and instrumentation and controls, additional benefit may be realized by consulting with contactors familiar with the owner and system. For example, if a specification for a supervisory control and data acquisition system uses a technically appropriate but high-end system, a knowledgeable contractor may recognize that a mid-range system can provide the same practical functionality and/or can be more easily integrated into the existing system.

7.3   Options Within the Design Process

In theory, a constructability review can be performed at any point in the design. In practical terms, a contractor can provide more specific and more valuable feedback when presented with a solid concept. Table 6.3 provides an overview of the areas a contractor can evaluate with each progressive stage of design. The later in a design phase a constructability review is conducted, the better the chances of receiving useful feedback. There is a risk of design rework if the constructability review identifies actionable flaws or challenges that the owner and engineer decide need addressing. While this situation does cost some time and money, it will be a faster and less expensive fix than if it is addressed during construction.

7.4   Construction Budget

Often, constructability reviews will ask the contractor reviewer if the project can be built within the construction budget established through design. This is especially important where state requirements or procurement methods prohibit awarding of projects outside the engineer's estimate of probable cost. This is a reasonable question, and experienced contractors will have a "gut feel" regarding whether the budget is adequate. The contractor reviewer likely will be able to point to aspects of the project with higher-than-normal costs. Sometimes this may be a physical constraint, such as groundwater or poor soils. Sometimes this will be a constraint that slows production, such as a single means of egress or overhead electrical wires. Sometimes this will be a sequencing constraint that lengthens project time, such as limitations on units out of service.

TABLE 6.3   Areas addressed by constructability review.

Design Phase	Areas Addressed
Preliminary engineering 10%–20%	General opinion on layout, phasing, challenges
30% design	Site layout planning, clearances required, long-lead equipment identification
60% design	Above, plus site/civil, mechanical, process, structural, maintenance of facility operations, sequencing
90% design	Above, plus electrical, HVAC
100% design	All the above

Generally, contractor reviewers do not perform a "real" estimate for a constructability review. This is because the level of effort to price in the same manner as performed for bidding requires soliciting pricing from subcontractors and vendors, who have no or little interest in a pricing exercise. For all parties, it is unlikely active bids will be put aside for a pricing exercise even if the contractor review is compensated because too much is invested in active bidding opportunities.

7.5   Considerations of Conflicts of Interest

Opinions regarding conflicts of interest in using contractor reviewers with the potential to bid or submit on a project may vary by state, and do vary by owner. Generally, doing a constructability review does not generate a conflict of interest because the feedback provided by the contractor reviewer benefits all future bidders/proposers equally. Contractor reviewers may have a small advantage in a bid as a result of having early exposure to the design, but this is a nominal advantage because the bid costs are driven by the cost of the work, vendor pricing, and subcontractor pricing, none of which are sensitive to the early preview.

Feedback of the following types may be considered inappropriate and should be scrutinized carefully before implementation:

- Recommendations to change to specific equipment vendors: Contractor reviewers may provide comments on installation, commissioning, or warranty experiences, but do not direct owners toward or away from engineer-selected equipment vendors.

- Recommendations to use proprietary services: Contractor reviewers do and should provide contacts for specialty services. The contractor reviewer should not provide nor direct owners toward subsidiaries or proprietary services that would strengthen their own position for the work.

7.6   Leveraging Results

As with value engineering, the usefulness of constructability reviews is determined by how the owner and engineer embrace the feedback. Neither owners nor engineers should view the comments as critical of the work done to date or the competency of other people or firms. Contractor reviewers are brought in because their expertise is in construction. The feedback provided must be evaluated against process design requirements, permitting requirements, and long-term operational goals. Saving money during construction may lower the bid price, but can set operations up for decades of difficult, inconvenient, and/or time-consuming work.

8.0   RISK IDENTIFICATION AND RISK REGISTERS

Risk identification is an activity that should be included in both value engineering and constructability reviews. This is a process of identifying those events that, if they happen, will interfere with the success of the project. When used in value engineering, events include those that depress the factors in the numerator of the value index, or inflate the denominator. In constructability reviews, these events inflate construction costs. The purpose of identifying risks is to evaluate whether they can be mitigated, deferred, or accepted. Mitigating risks means lowering either the likelihood the event will happen or the effect if the event does happen, or both. Deferring risks means putting them off to a future time, which might mean moving work to a different phase. Accepting risk means planning for the eventuality that the risk will occur by budgeting funds, people, or other resources to respond to that risk.

An internet search will yield many different templates to capture and analyze risks. These risk registers are tools to help all stakeholders understand risks. All risk registers break down to the following components:

- Identification of an event with consequence
- Description and/or analysis of the risk
- Rating of the probability/likelihood the event will occur
- Rating of the severity/impact of the event occurring
- Risk rating = probability × severity
- Identification of steps that can be taken to lessen/mitigate the probability or severity, or both
- Risk remaining after mitigation
- Final decision to accept, mitigate, or defer

Table 6.4 lists these components with an example to better illustrate their meaning.

General events can have a range of consequences, and the developer must use judgement when determining which events to include. Often, either the worst or most likely case is represented. In assessing probability, the developer should have some basis for judging the probability of occurrence. Instead of a probability percentage, probability of occurrence is frequently assigned on a category basis such as very low, low, medium, high, and very high. While this eliminates the need to assign a value, the developer still must have an understanding of the intent of the values covered by the categories.

Rating the severity of the impact can also create challenges because some events have financial impacts while others have human or environmental impacts. Here, again, the use of categories is common (low, medium, high)

TABLE 6.4 Basic risk register components with example.

Risk Register Category	Example
Identification of event with consequence	Known that a 2-in. gas line connects two buildings at a WRRF. Utility strike (event) during construction disrupts treatment process and injures workers (consequence).
Description of risk and analysis	As-builts show line as being outside the dig zone. Owner indicates as-builts are unreliable and has had problems on other projects with utilities not located where shown.
Probability of occurrence (1–5)	4: moderate because of history
Severity of occurrence (1–5)	5: high because of safety element
Risk	$4 \times 5 = 20$
Mitigating actions	1. Pothole in reported alignment to confirm location.
	2. Allow only hand-digging in area of concern. Affects contractors schedule and slows work, increasing costs for that portion of the project.
Residual risk	Gas line is found by shovel. Very low health and safety impacts. Very low process risks.

and the developer must have an awareness of the comparative rating across dissimilar impacts.

When identifying mitigating steps, it is best to identify those parties responsible for the mitigation. Mitigating steps may fall to those outside the design team or the contractor. Identifying the party responsible helps ensure the appropriate actions are taken.

Commonly, risk will remain after mitigation is taken. Including the adjusted probability and impact scores helps to ensure that all stakeholders recognize this fact. The process may also help identify other steps that can be taken to further mitigate risk.

Finally, the decision to accept, defer, or mitigate the risk should be clearly documented. Ambiguity increases risks. Identifying that risks are known with no clear approach to resolving them will only increase the questions and scrutiny should the event occur.

A word about costs. The list above does not mention costs. While some risk register formats do include costs, many do not. Putting costs to events helps the team understand what type of financial impacts can be incurred. But, the truth is that costs are difficult to estimate for many impacts. Hence, many of the formats use category surrogates for impact. Additionally, mitigating steps themselves are not without costs. More testing and investigation can significantly improve the understanding of the probability of an event occurring, but that benefit must be weighed against the cost that will be incurred for the additional investigation.

Other factors can be included in a risk register such as:

- Area impacts: schedule, cost, health and safety, other
- Schedule impacts
- Potential cost impact
- Responsibility: owner, engineer, contractor, other
- Discipline affected: geotech, civil, structural, process, electrical, other
- Cost of mitigation
- Priority rating

9.0   SUMMARY

Value engineering and constructability reviews are tools that owners can use to make good ideas great. This means maximizing the functionality, performance, quality, and operability of a system for each dollar invested. Value engineering and constructability reviews are not magical solutions for projects for which the scope is bigger than the budget. After the project concept and construction have been optimized, there may not be alternatives to cutting scope or increasing budget. The concepts of value engineering and constructability reviews are applicable to all projects of all sizes. The implementation of these tools is scaled, always staying in proportion to the size and complexity of the project. The end result of value engineering and constructability reviews is the confidence that the delivered facility is a glowing example of good stewardship of the ratepayers' investment and trust.

10.0   REFERENCES

Federal Acquisition Regulation Part 48. (2020). *Value Engineering.* https://www.acquisition.gov/content/part-48-value-engineering.

Federal Acquisition Regulation 52.248-1. (2020). *Value engineering.* https://www.acquisition.gov/content/52248-1-value-engineering.

Sherman, L. M. (2017, September 2). *Harvey's aftermath: Upwards movement for all commodity resin prices.* PT Online. https://www.ptonline.com/blog/postharveys-aftermath-upwards-movement-for-all-commodity-resins-prices-

7

Preliminary Design Phase

Tanush Wadhawan & Murthy Kasi

1.0   ELEMENTS OF THE DESIGN PHASE

The goal of the design phase is to prepare construction drawings and specifications that capture the concepts and intent of the facility plans. The design process described in Chapters 7 and 8 is intended to be applicable to all types of facility upgrades and expansions. For small projects, the owner and engineer may elect to simplify or combine some aspects of the process. The key is scaling the design effort to suit the construction activity. The owner and engineer may also agree to combine some of the phases or subphases to streamline the design process.

The design phase of a project requires engineering experience, expertise, and innovative approaches to develop design concepts, models, engineering drawings, specifications, and construction cost and schedule estimates, all in anticipation of the bidding and award of a construction project. Continuity between design and its preceding facility planning phases is vitally important. The amount of work that must be performed during the design phase, especially during the preliminary design phase, is directly related to the level of design performed during the facility planning phase.

This manual has subdivided the design work into two main phases: preliminary design and final design. Some engineering organizations may further subdivide these phases. In addition, the level of design associated with each of these phases may vary with the project delivery process of the specific engineering organization. The preliminary design phase may be anywhere from 15% to 30% of the design effort.

The preliminary design phase may be further divided into project definition (which can range from 5% to 15% design) and schematic or concept design (which can range from 15% to 30% design). The final design phase is typically considered to span 30% level design to 100% design as final completion, after that signed and sealed submission. The final design phase may be subdivided into detailed design, value engineering, and contract document preparation or final design (which is the completion of all design work). The design manuals of the various districts of the U.S. Army Corp of Engineers provide an example of this four-phase process and spell out, in detail, the requirements of each subphase (U.S. Army Corps of Engineers, 2008).

This chapter will address the preliminary design phase, while Chapter 8 will address the final design phase. Figure 7.1 depicts the phases and

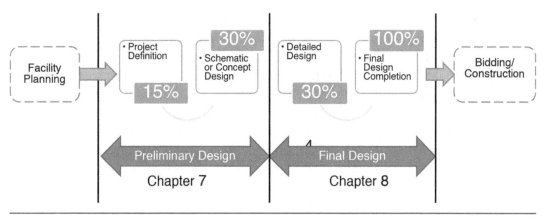

FIGURE 7.1   Elements of the design phase.

subphases of the design process and their relationships to the other parts of the overall process.

Chapters 7 and 8 are written primarily for the traditional design–bid–build process. For alternative delivery processes, such as design–build, the overall approach of these chapters is still generally applicable. Some of the details may vary to suit the specific delivery process. When a design–build delivery approach is used, the level of detail developed during preliminary design may be the same as, or may vary from, that developed for a traditional design–bid–build project. A key factor affecting the level of detail developed in a preliminary design effort for a design–build project is whether the project is delivered as a negotiated, qualifications-based project or as a competitive, cost-based selection. In a competitively bid design–build project, it is typical that the preliminary design level of detail is increased compared to that prepared for a traditional design–bid–build project. The reason for this is that the preliminary design package forms the basis for the design–build bid.

The level of detail during final design may decrease for a design-build project if some of the design elements are assigned to the contractors. For example, final lighting and communication system design may be assigned to the electrical contractor. The engineer only provides the basic information and criteria that must be met.

2.0   MANAGING THE PRELIMINARY DESIGN EFFORT

During the preliminary design, the concepts and vision developed during the planning phase of the project begin to take shape, and the project team builds the foundation for the detailed design of the facility. Work scope and activities should be focused during this phase of the project. The planning and evaluation of major alternatives was conducted during the facility planning phase of the project and should be refined but not be repeated here. One of the principal management concerns for this phase of the project should be to establish effective and efficient communication among all parties involved in the project. Project instructions or a work plan should be the first product developed and communicated in a preliminary design effort, or in any project for that matter. The work plan provides the tasks, schedules, and deliverables for the design team. If properly developed and endorsed by the engineer's design team, the owner, and other involved stakeholders, the work plan will lead to successful project completion. A quality management plan is an essential part of the work plan. This plan must outline the processes that will be embedded in the project by the engineer to ensure that the owner's expectations of quality are met or exceeded. Quality projects are the result of the project team being committed to producing quality work, and the team management being committed to managing quality. Quality is enhanced when tasks are completed correctly the first time. It is much more costly to correct mistakes at the time of final review than to eliminate them on a day-to-day basis.

The start of the predesign process is an excellent point in the project to bring all parties together for a chartering session. Chartering guides the team through the process of defining itself: its purpose, scope, goals, behaviors, roles, and responsibilities. Chartering sets expectations and ensures that the team members share the same vision for the project. It is a method of empowering team members and the group as a whole. The chartering session can be combined with the traditional project "kickoff" meeting. The session is generally attended by the owner's staff (management, operations and maintenance [O&M], finance, engineering, and others), the design engineering team, and other stakeholders (regulators, neighbors, and anyone with a vested interest in the project). Led by a facilitator, chartering consists of the following five-step process:

1. Defining the team—vision, purpose, boundaries, and linkages
2. Clarifying purpose—mission, priorities, success factors, and measures of success
3. Defining responsibilities—individual, team, and shared
4. Developing operating guidelines
5. Developing behavioral guidelines—core values, guiding principles, and conflict resolution

Once the elements of the charter are developed and recorded, the charter and work plan are endorsed by the team and all stakeholders. As the preliminary design progresses, the project team must communicate carefully with the owner to ensure that, as planning concepts take shape, these meet the owner's needs, preferences, and expectations—all while balancing the project budget and schedule. This is often where extra effort is needed to control excessive scope and budget "creep."

This process can be managed most effectively by a series of workshops where the owner's staff and the engineer meet to discuss specific details and issues about the proposed facilities. The workshops provide an excellent forum for face-to-face communication among the designers, operators, management, and other affected parties. Once the basic function and layout of the facilities are generally defined, issues such as equipment preferences, operating requirements, safety, and operation of the facility during construction can be discussed. This is a critical stage of the project, and management should focus on ensuring that all stakeholders in the project participate and clearly understand the objectives and the proposed solutions and cost implications.

As more detail about the project develops, issues such as zoning, codes, permitting, public outreach, legalities, and funding will broaden, and the number of agencies and stakeholders in the project will increase. The key to effective management here is to develop an effective and timely plan for proactively communicating and sharing information with these stakeholders. Control and management of this stage of the project can be with the engineer or owner (if adequate, experienced staff is available). Certain specialty areas such as funding and public outreach may be best handled by professionals who specialize in these areas.

At the conclusion of the preliminary design phase, a considerable engineering effort has begun. The preliminary design involves outlining design criteria, mass and energy balances, early development of process features and equipment selection, and process flow and instrumentation diagrams and narratives. Other design disciplines also enter the picture, including geotechnical and site investigations, electrical, civil and site, architectural, and building services. Because of the increased participation of additional engineering issues and disciplines, this project phase must be well planned, monitored, and controlled. In addition to the traditional management tools of workplans, budgets, and schedules, effective and frequent communication, properly documented, is the key to making the effort a success.

3.0   PRELIMINARY DESIGN OVERVIEW

The objective of the preliminary design phase is to further refine the process design that was selected during the facility planning process. It is critical that all major design decisions be made, endorsed (by the owner and the engineer's design team), and frozen by the completion of this phase. It is not expected that the preliminary design will change those design concepts established during the facility planning process. The amount of time and effort that is expended during this phase is highly dependent on the level of detail developed during the facility planning phase. Another factor that must be considered is the time lapse from the development of the facility plan to the start of the design phase. If there is a significant amount of time (more than 1 year) between completion of the facility plan to the start of the design, more time must be spent in revalidating the details of the facility plan. Another factor that may affect the level of effort is whether the same project team members (owner and engineer) who were involved in the facility plan will also be involved in the design phase of the project. Changing key project personnel (either owner or engineer) results in a loss of history regarding the basis for specific decisions. Initially, the new team may not fully agree with or understand the basis for the facility plan recommendations. Gaining endorsement of the plan or modifying it would become part of the preliminary design phase. Similarly, the owner's goals or financing situation may require revisiting the facility plan recommendations, as would changes in regulations.

The objective of the subsequent final design is to complete the documents required to obtain the construction bid and then build the project. Making changes in design decisions during the final design phase will require rework of the engineering performed during the preliminary design phase and will increase the overall engineering cost.

3.1   Drawings

Drawings (or sketches) prepared during preliminary design generally start out as rudimentary (in many cases, in freehand form), created by project engineers and architects to provide a simple document to communicate design concepts with the owner and other members of the design team. These preliminary drawings typically include either a process flow diagram or preliminary process and instrumentation diagram (P&ID), site plan, general building(s) plan(s) with basic equipment layout, typical building elevations, and at least one section through each facility that comprises the project.

Drawings and related documents (memos, etc.) created during the preliminary design phase are intended to provide the following information:

- Overall treatment process design concept and design criteria
- Basic site layout with major interconnecting conduits and pipes and grading, roadways, building locations, and any special civil or geotechnical areas delineated

- Basic building footprint and layout
- Major equipment listing with capacities, ratings, sizes, and utility requirements
- Equipment footprints and basic placement with a facility
- Maintenance requirements that will be incorporated to the design (space, access aisles, cranes and hoists, access, etc.)
- Walk areas
- Architectural and regulatory (building code, etc.) requirements, such as egress
- Room classifications (hazardous, explosive, corrosive, confined space, etc.)
- Safety considerations
- One-line electrical drawings (and need for standby power)
- Electrical room requirements (size, etc.)
- Utility requirements
- Mechanical equipment room size requirements
- Fire sprinkler requirements
- Basic instrumentation and control philosophy and P&IDs (starting as a process flow diagram with primary flow elements added; this will evolve to show secondary and interconnecting instrumentation and control)
- Chemical storage and feed requirements
- Equipment and structures to be demolished (but not demolition details)
- Cost estimate

For example, the drawings of a dewatering facility preliminary design would show placement of equipment, conveyors, walkways, ancillary equipment areas, electrical room, loading and unloading area, and building services requirements, at a minimum. When applicable, the drawings would show the location and size of the control room, truck loading area, mechanical room, polymer storage area, hoists or cranes, bathrooms, and pump rooms. The plan views and sections will clearly define dimensions of the building, including a preliminary height. It is expected that these drawings will be within ±10% of the final dimensions shown at the conclusion of final design.

During the preliminary design phase, these drawings are not expected to show structural components. This does not relieve the engineer of the responsibility of coordinating structural components during the layout of the facility. A preliminary determination of columns and beams and roofing components is essential. This coordination will ensure proper fit of equipment and enhance consideration of walk areas, equipment placement, and equipment support.

On occasion, upgrades are constructed in stages. When there is available information about the future upgrade or expansion, the space reserved for the equipment that may be added in the future should be identified. For example, if a new dewatering building is being constructed, with one centrifuge to be added now and one at some later date, the layout should indicate the potential layout footprint for the future centrifuge. Any anticipated ancillary processes required for the equipment that may be added in the future, such as piping, plumbing, containment curbs, electrical conduits, and spare input/output racks, should be noted on the preliminary drawings along with the basic areas allowed for these.

These preliminary drawings should also consider the paths for O&M personnel and the installation or removal of existing and future equipment. It is particularly important to consider how equipment will be added after the initial upgrade has been completed. For instance, if a centrifuge may be required in the future for a dewatering facility, a path through the facility to the space that has been allocated for this future addition should be provided.

3.2   Specifications

During this phase of design, equipment capacities, footprints, power requirements, and materials of construction are determined. This information is typically provided in an equipment summary table, with individual equipment data sheets for each piece of equipment. From these drawings and tables, an initial list of required specification sections is determined for use during the final design phase. This list would identify the major divisions of the specifications and identify the major process elements and units of equipment that need to be prepared during the final design phase. The specifications themselves are not prepared during preliminary design.

It is important to note that the equipment footprint, power, and other utility (seal water, drainage, etc.) requirements and ancillary system requirements vary (sometimes dramatically) by manufacturer and can have a major effect on the design and cost of the project. Because owners frequently have preferences on manufacturers of equipment, a meeting or discussion should be held with personnel representing both the operating and the maintenance organizations to determine owner preferences. This will allow the design to be based on those preferences. Frequently, regulatory and funding requirements dictate the need for a competitive bid and, unless otherwise directed in the specifications, the contractors may select a vendor other than the one used in the design. This can open the door to possible design changes during construction. Therefore, manufacturer selection is critical, both in a legal sense and to the design.

3.3   Design Standards and Automation Plan

A clear and easy-to-follow plan for design standards and automation is an essential element of project instructions, as it details the standards, processes, and procedures that will be used by the engineer to produce designs and drawings that meet project needs and the owner's documentation requirements. This plan addresses both the production of drawings and the electronic tools that the engineer will use in the design process. It is important that agreement between owner and engineer (and owner approval) be made upfront on the design automation approach because a change in design automation standards or procedures made in the middle of a project could prove costly. Owner requirements on computer-aided design drawings or design standards should be communicated before the project begins so that agreement may be reached on these issues. A more detailed discussion of the implication for software conflicts and data retention is provided in Chapter 8.

A decision should be made as to how existing facility data are to be collected and how the design team will use that information. Data collection can be costly or inexpensive, depending on the method of collection. The accuracy of the information will also vary greatly depending on the method and can affect the construction costs. The owner is generally aware of the accuracy of the existing drawings and is in the best position to guide the engineer on how to proceed in this area. Methods of data collection include:

- Use of record drawings or as-builts without field verification
- Field verification of record drawings or as-builts
- Traditional aboveground survey
- Underground probes and discrete excavation
- Three-dimensional laser scanning

Another issue affecting the cost of the design is how the design information and deliverables are to be used after design completion. This use may vary from the traditional copies of design contract documents to the use of the design information in the facility's life cycle (e.g., the hydraulic model used in the design of the facility may be used as part of a simulator for operator training). Chapter 8 has a more detailed discussion of the implications for the engineer's level of effort and potential legal implications.

A growing number of government entities, private organizations, and members of the design community have come to realize the importance of facility life cycle sustainability. There is tremendous pressure to provide and operate our infrastructure in an efficient and effective way.

Today, planning and design data can be created in such a way as to be sustainable and adaptable through construction and through operating the facility. Some of the most effective methods for creating sustainable life cycle designs involve the use of multidimensional (data-centric, object-oriented) automation tools to create "intelligent" three-dimensional models and schematics, coupled with an integrated database.

The evaluation of the design automation plan is best accomplished in a meeting that involves all stakeholders. Providing the basic information to all the participants before the meeting will facilitate a more meaningful discussion and allow participants to better prepare for the meeting. Recommended stakeholders include the owner's design automation staff and engineers, O&M staff, and the engineer's design automation specialist.

3.4   Elements of the Preliminary Design Phase

This section provides an overview of process- and design-related topics that are typically considered and refined during the preliminary design phase of a project. It is intended to finalize information presented in Chapter 4 so that it can be used in the final design of the treatment facility (Chapter 8).

3.4.1   Design Criteria

Establishing the final design criteria is perhaps the most important task in the design of unit operation and processes. The information is used to determine the number and size of structures and equipment. The design criteria provided in the facility plan should be reviewed and modified as required. The purpose is not to rewrite the facilities plan, but to achieve "buy-in" and continuity. This is particularly important when a different engineering firm is hired for the design phase or when design is initiated several years after the facility plan is finalized. It is important that this task is incorporated to the original scope of work. If it is included "after the fact" when funding is limited, the proper level of review and revision may not be possible. The following are some important considerations in the development of the design criteria:

- Influent flow and loads to a water resource recovery facility (WRRF) vary hourly, daily, and seasonally. These variations may be quantified by peaking factors, which should be determined from historic facility data. Facility design and plan recommendations should be reviewed to ensure that adequate capacity is provided to meet hydraulic and process needs under a range of conditions that are expected to be encountered during the design period. The range of conditions typically varies from a certain minimum in the initial years of operation to the maximum anticipated in the final years of the design period. Often, the minimum is overlooked, and the maximum is overstated, resulting in higher capital cost and lack of flexibility for optimized and cost-effective operation when actual future conditions depart from those expected (WEF, 2017). One idea to overcome this design problem, especially for facilities with a fairly long design period, is to install equipment sized for the initial period that can be easily replaced with higher capacity equipment as peak flows are reached to ensure operability throughout the facility's life. Factors that affect future influent characteristics, such as potential reduction in infiltration and inflow, industrial and residential growth in the service area, and water conservation, should be assessed as accurately as possible. Table 7.1 provides typical flows and mass loading factors used in WRRF design.

- Sidestreams originating from sludge-processing operations can impose additional and significant nutrient and solids loading on the mainstream process. Although recycle loads affect the performance of all WRRFs, the biological nutrient removal process is particularly vulnerable. Facilities that operate well below the design capacity may not see symptoms of recycle overloads because of the availability of adequate treatment capacity. While recycle overload may not be an issue initially, operational costs to treat the recycle can become an issue over time; therefore, the design should consider how to minimize such costs. Nonetheless, mitigation measures should be planned for phased implementation as the facility approaches design capacity. The adverse effects of return streams can be eliminated or controlled by proper planning, design, and operation (Qasim, 1999).

- As indicated in Chapter 4, the required degree of treatment should be established considering the influent characteristics and the target effluent limits. Typically, the target effluent limits are lower than the permitted limits to account for operational variability and process reliability. Process design should also include appropriate safety factors to overcome unknowns and unusual operating conditions. This is particularly true if stringent effluent limits must be met. Because regulations are constantly evolving, facility design should incorporate adequate flexibility to respond to anticipated future regulatory changes.

TABLE 7.1   Effect of flows and loads on WRRF design and operations (adapted from WEF [2017] and Metcalf & Eddy, Inc. [2014]).

Factor	Condition	Typical Purpose
Flow rate		
	Average annual day	Determining peaking factors and O&M costs.
	Minimum hour	Pump turndown requirements; flow meter low range.
	Minimum day	Size conveyance system. Determine mixing requirements and trickling filter recycle rates.
	Minimum month	Determine number of units required during low flow condition. Schedule shutdown for maintenance.
	Peak hour	Determine pump capacity and size conveyance system. Size preliminary and primary unit operations, final clarifiers, filters, and chlorine contact tanks. Develop strategies for managing high flows.
	Maximum day	Size equalization basins, chlorine contact tanks, and sludge pumping systems.
Mass loading		
	Minimum month	Determine process turndown requirements. Check biological nutrient removal system performance under weak loading conditions in conjunction with anticipated recycle loads. Check alkalinity availability (typically, high rainfall month) for nitrification with anticipated recycle loads.
	Minimum day	Determine biological system recycle rates.
	Maximum day	Check ability of aeration system to maintain a slightly reduced dissolved oxygen of approximately 1.0 mg/L.
	Maximum month	Design biological processes, aeration system, and sludge storage facilities.
	7-day minimum	Check alkalinity availability (typically, high rainfall period) for nitrification with anticipated recycle loads.
	15-day maximum	Design sludge stabilization processes.

- Typical design criteria for unit processes that compose the treatment train are published in various documents (see References and Suggested Readings). Facility operational data should be reviewed to determine if there is a need to revise typical design criteria to suit site-specific conditions (e.g., the final clarifier design criteria may need to be modified to accommodate higher-than-typical sludge volume indices). If pilot- or demonstration-scale test results are available, these should be used as guides to establish site-specific design criteria.

It is important to recognize that reduced flows and loads during the early years of operation before the facility nears the rated capacity can cause operating difficulties. The following examples illustrate this point:

- Reduced flows can cause less-than-adequate velocities in pipes or channels and result in solids settlement.
- In enhanced biological phosphorus removal facilities, excessive anoxic or anaerobic retention times caused by reduced flows may trigger secondary phosphorus release and compromise process efficiency.
- Certain types of chemical feed pumps have a narrow operating range and are unable to accurately meter fluids below a certain percentage of their maximum capacity.
- Aeration systems may be oversized and may deliver excessive air during periods of low loading, such as in the early morning hours when flow and load are typically at the lowest level. This may cause oxygen bleed from the aerobic zone to the downstream anoxic zone (second state) and result in reduced denitrification performance.

Consideration should be given to providing duplicate units of equipment or structures so that some of the units can be taken out of service during reduced flow periods.

3.4.2 Performance Criteria

While the design criteria are used to size structures and equipment, the performance criteria are linked to system operation. For example, shutting down one primary clarifier may overload the remaining units, such that the performance criteria (solids-removal efficiency) cannot be met. That is, no matter how well designed a treatment unit is, performance criteria will be compromised if it is operated in a manner such that the design criteria are not met.

The performance criteria presented in the facility plan should be reviewed and revised, if required. Typical performance standards for commonly used unit processes are available for comparison (refer to References and Suggested Readings at the end of this chapter). Site-specific data from pilot- or demonstration-scale studies, if available, should be considered in specifying performance criteria.

3.4.3 Reliability and Redundancy

"Reliability" and "redundancy" are interrelated terms with different meanings. Reliability of equipment refers to its dependability in achieving the design objective, while redundancy refers to the provision of additional equipment to improve process reliability. Maximum reliability is achieved when no single component failure at full design conditions can impair the performance of any individual unit or the facility as a whole. System reliability increases with an increase in redundancy by providing additional equipment and reserve capacity, such as "firm" capacity; however, the capital costs also increase. Firm capacity is the capacity of each unit operating with the largest unit out of service.

Often, because of the high cost of providing complete reliability (i.e., meeting effluent limits at all times) under design conditions, utilities in some states have chosen to define an acceptable level of treatment in terms of percent compliance. For example, five ammonia nitrogen (NH_3-N) exceedances of the daily maximum (not-to-exceed) limit per year represents 98.6% compliance. It should be noted that, often, the National Pollutant Discharge Elimination System (NPDES) permit spells out certain maximums that cannot be exceeded. If it is not included in the permit language, the owner and engineer should jointly establish the "acceptable risk." Coordination with local regulatory agencies would allow this determination to be made with full knowledge of the potential consequences. Owners and engineers should approach the subject of acceptable risk with caution, fully weighing the cost of noncompliance, public health, environmental impacts, reclaimed water quality, and public perception. For example, while the number of exceedances per year may not result in regulatory penalties, the resulting poor reclaimed water quality might foul the cooling equipment of a reclaimed water customer's industrial process.

Reliability and redundancy requirements are linked to overall performance variability, which is composed of the following:

- Influent flow and load variability: As stated previously, influent characteristics exhibit diurnal, weekly, and seasonal variations.
- Process variability: This is caused by inherent variability associated with the chemical, physical, and biological process reactions and external variability related to facility design and operation.
- Equipment reliability: This refers to variability of the mechanical equipment used at the WRRF. A discussion of approaches available

for determining mechanical reliability may be found in U.S. Environmental Protection Agency (1982) and Metcalf & Eddy, Inc. (2014).

Facility reliability is achieved through standby equipment, multiple units,
and an overall system providing multiple opportunities to achieve performance objectives (such as firm capacity). Reliability can be incorporated
into a facility by a separate analysis of risks, costs, and benefits. Alternatively, generally accepted criteria, such as the U.S. Environmental Protection
Agency (U.S. EPA) reliability classification (Class I, II, or III) and sufficient
peak capacity may be used (Corbitt, 1998). In addition, many local regu-
lations/standards/criteria specify redundancy requirements for commonly
used wastewater treatment processes and support systems. Finally, process
modeling can be a powerful tool that will help identify the inherent system
reliability through various combinations of equipment failures and subsequent operational changes required to meet the effluent permit limits
(Andraka, 2020).

3.4.4   Develop or Refine Mass and Energy Balances

The mass balance is the primary tool for understanding the treatment process by quantifying the changes that occur to wastewater constituents during
treatment. The mass balance concept is based on the principle that, although
the constituents may change form, mass is conserved. If a mass balance was
developed during facility planning, it should be evaluated and refined if
needed; if not, then one should be developed during the early stages of the
preliminary design. Mass balances are developed using actual operating data
and should include all inputs and outputs, including recycle loads. Typically,
average conditions are considered and the mass balance is adjusted to assess
the peak conditions and loading conditions with units out of service. Both
present and future scenarios should be examined.

Development and use of steady-state models or spreadsheet-based models to assess the mass balance around the critical processes and perform preliminary design calculations is highly recommended. For instance, a solids
mass balance around solid-liquid phase separators in a facility using actual
data can be used to assess the quality of facility data and evaluate solids
treatment capacity from a design point of view. This can be particularly
useful in resolving issues with model calibration and/or validation.

3.4.5   Develop Process Models

Process models are tools that allow the designers to ask "what if" questions
and explore system behavior under dynamic and steady-state conditions.

Although they provide value to the designer, they are not central to the design process. Process models are not always required as part of the design process.

Early process modeling and simulation efforts focused on steady-state behavior of the wastewater treatment process. Modern computer resources, however, have made dynamic models practical. Commercially available process modeling software include Sumo, Biowin, GPS-X, Simba, and WEST. The mechanistic model developed by the International Association on Water Quality is one of the more readily used models. Mechanistic models are based on fundamental principles of physics, chemistry, and biology. These robust models are applicable over a wide range of operating conditions. Model development typically entails the following three steps:

1. Review of facility layout and operating data
2. Dynamic model construction
3. Model calibration and validation

Model development relies heavily on significant reliable process operation and performance data. Often, it requires collecting additional data on influent characteristics not typically measured (e.g., constituent fractionation) because of the complexity of the new models available and the importance of calibrating these models under a variety of load and operating conditions. The goal of model calibration is to minimize the difference between model output and actual data. This is an important part of model development because the model reliability is a function of the quality of the data used in the calibration. The amount of data required for model calibration depends on the type of model. Steady-state calibration can use existing daily, weekly, or monthly averages that are obtained from grab or composite samples. Dynamic calibration relies on instantaneous values obtained from online analyzers or grab samples. The type of data required fall into the following four categories:

1. Physical data (process schematic, bioreactor flow pattern, tank dimensions, and volumes)
2. Operational data (process variables and effluent characteristics)
3. Influent characteristics
4. Kinetic and stochiometric data

The calibrated model is validated with a separate set of data to ensure the use of the calibrated model with the level of confidence required to meet modeling objectives, such as whether the calibrated model is used for process design or comparison of alternatives, or both. The level of effort required for this process depends on the application and the corresponding level of detail expected from model simulation results.

One of the most important uses of process models is that they can be used in a process train audit to answer "what if" questions. On-site tests and measurements are typically completed in conjunction with the modeling effort. This provides insight to the process from the model output, and the model can be improved by observing facility performance.

Dynamic process model evaluations of a facility can be especially important to determine design reliability by accounting for influent flow, load variability, and process variability. A dynamic simulation can help identify the unknowns that cannot be obtained from a steady-state spreadsheet-based calculation. As the demand for nutrient removal and recovery is increasing, the usefulness of modeling platforms in performing preliminary and final design cannot be ignored. This is because biological nutrient removal processes behave differently when they are repeatedly subject to variable influent flow and load conditions.

Diurnal flow simulation is the simplest dynamic simulation to evaluate a system's performance reliability when exposed to a daily repeated typical flow and load variations. An extension of diurnal modeling is to simulate system response to weekly or monthly, or even seasonal, variations. The most important dynamic simulation analysis to run for design evaluation and system sizing is the "birthday cake" analysis. This is performed by considering design loadings and including annual average, maximum monthly average, and peak day events. If these data are unavailable, an artificial loading pattern with influent loads increasing and decreasing with time, resembling the shape of a tiered cake, can be used as input to the model, as shown in Figure 7.2. For further details on dynamic model evaluation using birthday cake analysis, refer to *Wastewater Treatment Process Modeling* (WEF, 2014).

4.0 MAINTENANCE OPERATIONS DURING CONSTRUCTION

During the preliminary design phase, it is important to refine and solidify the information from the facility planning phase on how existing facility operations function and how functionality will be affected and managed during the construction. At this phase, the engineer and facility operations staff must work closely together—preferably in a series of workshops held at the facility—to evaluate critical processes. At a minimum, the following items

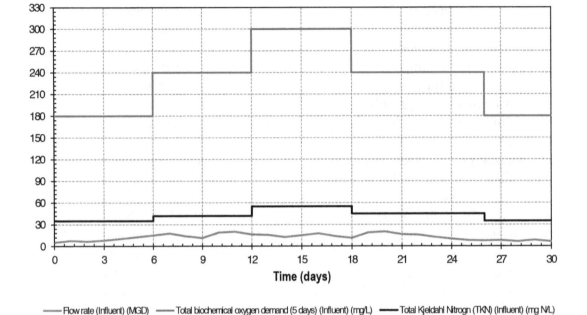

FIGURE 7.2   Example dynamic input resembling "birthday cake."

must be carefully considered and evaluated during the preliminary design phase and before final design and development of the contract documents:

- Which unit processes can be taken offline and for how long a period?
  - Will taking a unit process out of service generate a need for an interim permit?
  - Will the efficiency of some treatment units need to be enhanced to make up for the loss of a unit during construction? For example, will polymer addition be required in secondary clarifiers to make up for one unit being taken offline for upgrade?
  - What are the effects of shutdown on reclaimed water customers?
  - What are the NPDES requirements to report and control construction impacts?
  - What are the effects of shutdown on the biological processes, including activated sludge and anaerobic digestion?
  - Will seed to start up the process be necessary? If so, is it readily available?
- Can the facility be completely shut down for periods of time to allow major piping connections or equipment or utilities removal or installation?

○ For how long?

○ How often?

○ Must it be done at low flow or under other special conditions or restraints?

• Can process flows be turned off for various periods? If so, how long without degrading the process?

The contractor must be aware of these constraints when bidding on the project. When developing the construction documents, even in the preliminary stage, it is essential to consider and include these constraints within the design documents.

During the preliminary design, the engineer must continue to evaluate and refine the constructability of the improvements and consider the sequence of construction as an integral part of the design to confirm that the facility process necessities can be adequately maintained during the construction. Consideration can then be given to the initial scheduling and sequencing of existing unit process upgrades, facility shutdowns, and process flow interruptions.

Take, for example, an existing facility having three secondary clarifiers. The facility requires, at a minimum, two clarifiers in operation. The design calls for renovating the existing clarifiers and constructing a fourth clarifier. Because of process piping limitations, the contractor must take two clarifiers offline at any one time. Therefore, the design documents must reflect the fact that the fourth clarifier must be built and online before any renovation so that the proper desired clarification capacity is maintained throughout the construction period.

During this preliminary design phase, the engineer should become familiar with existing facility operations so construction and design can be tailored around operations. The engineer should also consult with facility personnel about how to maintain the most efficient facility operation during construction. Together, they need to look at how to minimize disruption by considering such things as bypass pumping to convey flow from one process to another while new pipe is being installed, specifically coordinated construction sequencing, alternate temporary piping routes, and specially designed temporary facilities.

Depending on the size and duration of the construction upgrade, special consideration must be given to the sequence as it relates to new facilities being put into the process stream. For example, if unit processes will be completed and placed into service at various times throughout the construction period and final control systems (e.g., supervisory control and data acquisition systems) will not be completed until later, interim controls and alarm systems may be needed to ensure that the process can be functionally operated and protected until such time as all the related components are complete. Depending on the type of facility upgrade, it may be best to put new facilities into service as early as possible for two important reasons: first, the contractor is on-site and can provide additional support as it relates to equipment function and, second, the contractor will be on-site to service any warranty issues. During this period, facility operators can become accustomed to and comfortable with the new facilities.

5.0   INTERIM PERMIT REQUIREMENTS

During the preliminary design phase, the engineer should determine if the existing facility would be able to maintain permit requirements throughout the construction phase. The engineer should take into account which unit processes will be affected, and for how long. If the upgrade will decrease biological reactor or clarifier volume, then there will be a need for interim biochemical oxygen demand, total suspended solids and, possibly, ammonia limits. The owner must request these interim permit limits from the regulatory agency and give that agency enough time to evaluate that request. Some regulatory agencies will not allow interim permit limits for some effluent parameters, and the engineers must then sequence construction in such a way as to avoid permit violations. When requesting interim permit limits, the engineer should provide justification to the owner for the requested limit. The justification must be based on sound technical judgment. It is always best if the owner and engineer communicate early with the regulators before formally submitting any variance or interim request.

There are other facility operations conditions that do not affect the process, but do affect personnel and their ability to operate and maintain the facility. Items such as temporary roadways, walkways, and handrails all need to be considered during design. Other considerations include:

- Sufficient parking for facility and construction staff
- Deliveries of equipment and supplies
- Construction material storage areas and access to buildings or tanks
- Location of construction trailers
- Temporary laboratory or office space
- Safety of visitors, vendors, and staff

If the engineer begins early in the design process to build on communications with the owner and staff and information developed in the facility planning phase to consider these details, there is a much higher probability that the project will proceed as anticipated and be designed with minimum contractor claims stemming from process-related maladies.

6.0 OTHER CONSTRAINTS, REQUIREMENTS, AND REGULATIONS TO CONSIDER DURING PRELIMINARY DESIGN

There are many potential constraints, requirements, and regulations for these types of projects. The majority of items discussed are site-, community-, region-, and state-specific, so it is important that these issues are investigated thoroughly for each individual case. If careful attention is not paid to these issues, there can be significant and costly delays or it may be necessary to redesign certain aspects of the upgrade, which can substantially increase costs and affect the schedule. Obtaining land or easements also requires many of the factors listed in this section as well as probable environmental, legal, and public input.

This section discusses many of the codes and permits that are required for the upgrade. Although this section focuses on government regulations and their required permits, some aspects also apply to utility connections. All team members need to know who is responsible for what; what kinds of codes, regulations, and utility requirements need to be complied with; safety and personnel requirements; and why it is important to evaluate site history. The discussion that follows will offer guidance about these issues. In all cases, proactive communication between the project team and regulators and utilities is important. All information, including copies of applicable codes, regulations, and other data, should be thoroughly documented.

6.1 Responsible Party

It is important to establish at the start of the project the party responsible for investigating each constraint, regulation, or requirement. The owner and engineer should establish the basics of who will apply for permits in the contract's scope of work. In some cases, it may be obvious from either state or local regulations. Some communities will require the engineer to obtain any permits from regulatory agencies, while others may want to do it themselves and, in some cases, it will be a contractor's responsibility. It is important to have a written schedule indicating the responsible party and when the permit must be in place. This should be done early in the preliminary design phase. Development of a checklist can help ensure that permits have been obtained. Information that should be in the checklist include the type of permit, the person or agency responsible, the length of time required to obtain it, when it is needed (before or during design, before or during construction), and whether it has been obtained. The list should be updated frequently and reflect new or changed permitting responsibility.

6.2 Zoning

Almost every community will have zoning regulations administered by a governing board. Even though this is an upgrade of an existing facility, zoning requirements may have changed since the facility was originally constructed. During this planning phase, the owner and engineer should meet with the zoning officials to present an overview of the project and determine if there are any features that may not meet zoning regulations, or require a variance. The next step is for the owner and engineer to decide who will apply for the zoning permit, and when. Zoning regulations may require installation of a decorative wall or extensive landscaping, dictate a particular architectural style or building height, or require buffer zones. They may also include coastal area management requirements and floodplain analysis. Zoning regulations can significantly affect the cost and final design of the project. Zoning requirements investigated and identified during the facility plan need to be finalized early in the design phase.

6.3 Flood Zone and Floodplain Analysis

During floodplain analysis, regulations of state, province, and federal agencies must be considered. The Ten States Standards recommend that all pumping stations and treatment facilities be fully functional and accessible during the 25-year flood, while the equipment, electrical, and structures must be protected from a 100-year flood event (Wastewater Committee of the Great Lakes-Upper Mississippi River, 2014). Flood maps can be used to determine whether the selected site for treatment facility construction is in a 100-year flood zone, a 500-year flood zone, or outside of a 500-year flood zone. If the facility will be in one of the flood zones, vulnerable equipment will be identified along with alternative methods to protect the equipment from flooding. Although the facility may be located within a floodplain, the equipment may not be vulnerable to flooding if it is built or located on an elevated platform or within a bermed area.

6.4 Codes and Regulations

There are many codes that affect an upgrade project. These codes vary by community, county, and state, and must be investigated and fully documented. The responsible party for this should be the design engineer, and these must be determined early in the preliminary design phase. In addition

to building, accessibility, and safety codes, there will be codes for maintenance holes, catch basins, paving, electrical installations, plumbing, and so
on. The owner must ensure that the design engineer has investigated all
applicable codes before designing the facility.

6.5 Building Codes

There are various building codes that govern much of the upgrade design.
The most common building codes are listed in Table 7.2. It is critical that the
designers investigate all codes because many will be site- and region-specific,
such as for hurricane- or earthquake-prone areas and areas that can expect
large amounts of snow. Each state may have an individual building code
that might supersede all the other building codes. It is important to have a
detailed code check performed by all appropriate team members (architectural; process engineering; structural; electrical; plumbing; heating, ventilation, and air conditioning (HVAC); and civil engineering representatives) as
one of the first steps in the preliminary design phase.

TABLE 7.2. List of some common building codes.

Code	Purpose
International Building Code and Building Officials and Code Administrators International (BOCA), National Building Code	Covers all aspects of building, including structural, mechanical, and electrical construction
International Code Council (ICC) codes	Covers mechanical and electrical construction
Individual states building codes	Covers items that might have stricter compliance requirements from the BOCA or ICC codes
American Society for Testing and Materials	Covers materials of construction and building codes installation
National Electrical Codes	Covers requirements for electrical installation
National Fire Protection Association	Covers electrical installation requirements specific to wastewater treatment facilities
International and Uniform Fire Codes	Covers fireproofing requirements
American Society of Heating, Refrigeration and Air Conditioning Engineers	Referred to in many building codes for heating, ventilating, and air conditioning design and construction requirements

6.6   Permitting Requirements

Many different types of permits will be required for an upgrade project. These include permits issued by local, state, and even federal agencies. One of the most important is the NPDES permit for the facility. It is important to negotiate the final version of this permit with the regulatory agencies as early as possible to ensure that the design will meet current and future requirements.

Other types of permits include the following:

- Building permits
- Coastal area management permits
- Disruption permits
- Demolition permits
- Dewatering permits
- Local environmental protection permits
- Stormwater management permits for discharges from impervious surfaces
- Soil and erosion control permits
- Permits that allow for reduced effluent standards during construction
- Other local and site-specific permits such as inland wetlands, local historical commission, and U.S. Army Corps of Engineers

WRRFs are considered Sector T industrial sites under the General or Multi-Sector Storm Water Permits, and thus have sampling requirements for stormwater outfalls. The number of outfalls should be minimized to reduce operational effort. The outfall locations should be accessible in a safe, nonhazardous location because well sampling may be required during storms and at night.

Many communities waive the cost of building permits for municipal projects. If this is the case, this information should be conveyed to the contractors at the time of the pre-bid conference. It is important for the design engineers to identify all required permits. The party responsible for obtaining these permits must be clearly shown in the contract documents and specifications. Table 7.3 is an example of a checklist that can be used to ensure that the permits are applied for and received so that the project is not delayed.

It is important for the engineer, construction contractor, and owner to understand who is responsible for what. If permits are not obtained in a timely manner, it may delay construction and result in added expense to the municipality.

TABLE 7.3 Checklist of various permits.

Permit	Time Need to Obtain	Permitting Authority	When Needed[a]
National Pollutant Discharge Elimination System	Owner/engineer	State regulatory agency	As early in design as possible[b]
Land use/zoning permit	Owner/engineer	Local agency	During final design[b]
Coastal area management and flood	Owner/engineer	Local agency	During final design[b]
Disruption permits	Engineer or contractor	State solid waste department	Before the start of excavation
Building permits	Contractor	Local building department	At start of project
Demolition	Contractor	Local building department	Before demolition
Air permit	Owner/engineer	State regulatory agency	Before construction
Dredging or filling navigable water	Owner/engineer	U.S. Army Corps of Engineers	Before construction
Dewatering/surface discharge permits	Contractor	State/federal regulatory agency	Before excavation

[a]The substantial requirements for all permits should preferably be understood as early in the project as possible to minimize "surprises."
[b]The requirements for the permits noted must be obtained and used in the project evaluation for the facilities plan.

6.7   Legal Issues

The owner should designate an attorney (staff or outside) who will be responsible for exploring any legal requirements and for reviewing contract documents, including general conditions and the actual construction contract before this project being bid. There should also be a legal review for any property acquisitions, zoning variances, permit negotiations, and so on.

6.8   Funding and Financing

Availability of funds for upgrades and expansions to treatment facilities have been dwindling over the years as other demands take priority for federal, state, and local governments. Both the Water Environment Federation and the National Association of Clean Water Agencies are working diligently to support increases in federal funding of water and wastewater infrastructure needs.

6.8.1 State or Federal Funding

Often, funding for upgrades is through Clean Water Act programs. These can be grants, low-interest loans, or combinations of the two. In addition, there may be special programs such as the Long Island Sound improvement program in Connecticut, which gives a 30% grant for any nitrogen-removing facility upgrades. It is important to investigate all funding sources. Many states have priority lists, and funding priority is associated with placement on the priority list. It is important to ensure that all information concerning the need for an upgrade is transmitted to the regulatory agencies. In addition, some states use the U.S. EPA's Computer Assisted Procedure for Design and Evaluation of Wastewater Treatment Systems (CAPDET) funding formula. It is important to investigate whether your state uses this formula and how it will affect upgrade funding.

6.8.2 Local Funding

Local funding is typically through general obligation bonds or revenue bonds. Each community will have a different safe debt limit. It is important to plan these projects and time their construction to conform to any safe debt limits imposed by rating agencies and the community's finance directors.

Each community will have its own requirements for acceptance and approval of funds. It is important to understand that process and begin it early. For example, you may need to receive approval from the local planning board, wastewater authority board, board of finance, or board of representatives. It is also important to meet with the community because it will be paying for the project, either directly or indirectly. For example, the Stamford Water Pollution Control Authority (SWPCA) in Stamford, Connecticut, in preparation for an $105 million upgrade and expansion project, produced a 20-year, pro forma spreadsheet showing the effect of the debt service on the user charge. This became a great tool for explaining to the public about the increases they would see, especially for the first 5 years. The SPWCA made a special effort to level the increases and avoid increases greater than 8% in any fiscal year. With this understanding, funding for the project was approved unanimously, even though this was the largest single capital project ever undertaken by the city of Stamford. An example of that spreadsheet is shown in Table 7.4.

TABLE 7.4 Pro-forma user charge spreadsheet.

	FY 03/04	FY 04/05	FY 05/06	FY 06/07	FY 07/08	FY 08/09
			Adopted Plan			
Revenues						
Net user fees	$10,125,035	$10,837,668	$11,424,162	$12,189,990	$12,531,700	$12,586,422
Delinquent sewer user fees	$550,000	$548,625	$547,253	$545,885	$544,521	$543,159
Interest and penalties	$92,284	$97,360	$98,684	$116,940	$115,537	$116,993
Sewer assessments	$1,100,000	$1,135,000	$1,135,000	$1,135,000	$1,135,000	$1,135,000
Darien wastewater charges	$1,039,726	$1,060,521	$1,081,731	$1,103,366	$1,125,433	$1,147,942
Septic tank dumping fees	$205,821	$203,763	$201,725	$199,708	$197,711	$195,734
Regional laboratory fees	$90,000	$90,450	$90,902	$91,357	$97,450	$97,938
Darien capital contribution	$115,000	$115,000	$115,000	$115,000	$115,000	$115,000
Darien capital contribution—upgrade				$979,446	$534,930	$534,930
Interest income	$15,000	$45,000	$80,000	$50,000	$45,000	$44,000
Nitrogen credits trading income	$150,000	$234,500	$323,200	$441,500	$350,000	$300,000
Use of rate stabilization reserve				$3,400,000	$1,000,000	
Total revenues	$13,482,866	$14,367,887	$15,097,657	$20,368,192	$17,792,282	$16,817,118

(Continued next page)

6.9   Schedule

In many cases, upgrades are part of a consent order or other legal requirement imposed by the regulatory agencies, and these agencies dictate the compliance schedule. Building on the facility planning phase information, in the design phase it is important to understand the constraints of the compliance schedule and evaluate contractibility based on that schedule.

TABLE 7.4 Pro-forma user charge spreadsheet (*continued from previous page*)

	FY 03/04	FY 04/05	FY 05/06	FY 06/07	FY 07/08	FY 08/09
			Adopted Plan			
Operating expenses						
Labor	$2,432,952	$2,505,941	$2,581,119	$2,658,552	$2,738,309	$2,820,458
Employee benefits	$1,011,036	$1,084,849	$1,164,985	$1,252,035	$1,346,641	$1,449,511
Maintenance	$568,775	$504,130	$385,422	$387,445	$389,479	$391,526
Supplies and other expenses	$1,069,367	$1,094,721	$1,125,332	$1,156,859	$1,189,329	$1,222,772
Sludge haul away	$1,250,040	$1,437,546	$1,466,297	$1,495,623	$1,400,000	$250,000
Utilities	$1,000,934	$1,032,073	$1,064,249	$1,097,501	$1,130,189	$1,163,854
Contingency	$180,000	$300,000	$300,000	$300,000	$300,000	$300,000
Reserve for state Clean Water Fund borrowing	$1,836,000	$1,154,246				
Rate stabilization set aside		$1,275,000	$3,125,000			
Total operating expenses	$9,349,104	$10,388,506	$11,212,404	$8,348,015	$8,493,947	$7,598,121
Net revenues available for debt service	$4,133,762	$3,979,381	$3,885,253	$12,020,177	$9,298,335	$9,218,997
Debt service						
Senior lien revenue bonds (projected)	$360,000	$800,000	$800,000	$800,000	$1,970,000	$1,970,000
Senior lien—state of Connecticut outstanding	$436,006	$432,410	$428,815	$425,219	$421,624	$418,028

(Continued next page)

Weather conditions and other seasonal requirements will also affect the schedule. The design and specifications need to take into account the effects of cold or wet weather on meeting scheduled compliance dates.

If the project is the result of a consent decree or other regulatory order, there will be a compliance schedule that must be met. It is good practice to keep regulators informed about the progress, either through regular meetings or interim compliance reports.

TABLE 7.4   Pro-forma user charge spreadsheet (*continued from previous page*)

	FY 03/04	FY 04/05	FY 05/06	FY 06/07	FY 07/08	FY 08/09
				Adopted Plan		
Debt service (*continued*)						
Senior lien—state of Connecticut planned				$8,162,047	$4,457,746	$4,457,746
Senior lien debt service coverage	5.193	3.229	3.162	1.280	1.358	1.347
Subordinated lien—city of Stamford G.O. bonds	$3,030,750	$2,746,971	$2,656,438	$2,632,911	$2,448,997	$2,362,098
Total debt service coverage	1.08	1.0	1.0	1.0	1.0	1.001
User fee calculation						
Uncollected reserve (5%)	$532,897	$570,404	$601,272	$641,578	$659,563	$662,443
Gross user fees	$10,657,932	$11,408,072	$12,025,434	$12,831,568	$13,191,263	$13,248,865
Annual consumption (100 cu ft)	$5,169,022	$5,194,867	$5,220,841	$5,246,946	$5,273,180	$5,299,546
Annual increase in consumption, %	0.44%	0.50%	0.50%	0.50%	0.50%	0.50%
User charge (rate) per 100 cu ft (FY00/01=1.65)	2.06	2.2	2.3	2.45	2.5	2.5
Annual percent increase	7.88%	6.70%	6.86%	6.52%	2.04%	0.00%

6.10   Standards and Preferences

During the planning phase of the preliminary design, the owner and engineer must finalize standards and preferences to be used for the project. The use of many design standards is dictated by the regulatory agency. The design standards will address such things as detention times and overflow rates for sedimentation processes, length-to-width and width-to-depth ratios for grit chambers and rectangular sedimentation tanks, and volatile solids loading rates for digesters, to name a few. It is important to note that these "standards" are generally considered "guidelines" that are good practice to follow but may not be appropriate for all situations. Furthermore, some communities may have their own design standards that they want to use. Typically, designers use a combination of various standards based on their experience and successes with other upgrades and input from the owner.

In addition to design standards, it is important to establish standards for data automation (see Section 3.3 in this chapter), equipment and instrumentation, and architectural style.

6.11 Other Standards

Owners or the city or town engineers may have specific requirements for materials of construction, equipment manufacturers, and equipment types based on regulations and their experience. The engineer needs to respect those requirements yet still meet any competitive bidding requirements. In addition, there are standards for contracts and progress payments, which are discussed in detail in Chapter 8.

6.12 Architectural Styles and Landscaping

For an upgrade, the owner typically wants to match existing architecture if it is satisfactory. The architectural style must also meet zoning requirements, building codes, and public acceptance. In most cases, it is relatively easy to match existing architecture; however, in some cases, it might add significantly to costs. In that case, the owner and engineer or architect need to adopt the building architecture that does not look "out of place" but is more cost effective. The goal should be that the treatment facility looks nice. The residents of a community generally only see and appreciate the cosmetic aspects of the site. If it looks good and pleasing to their eyes, they assume that it is running well. If it looks poor, then no matter how well run it is, the residents will think that it is poorly run. Their first, and sometimes only, impression is with the buildings and grounds. It is important that attention is paid to these areas.

Certainly, landscape design is influenced by climate and how complex maintenance will be. Heavy investment in high-maintenance landscaping should be a conscious decision by the owner, considering long-term cost, reasons for landscaping, effect on neighbors, and odds of the vegetation surviving as designed. Low-cost and environmentally sound options such as xeriscape, prairie, and drought-resistant species should be considered, based on climate and zone.

6.13 Safety Requirements

The WRRF environment contains many potential biological, chemical, and physical hazards as a result of the nature of wastewater and byproducts generated, and the treatment processes and equipment used. Approximately 20% of accidents are directly attributed to physical conditions, many of which can be prevented during design. The other 80%, caused by work methods, can be reduced indirectly by design changes that avoid hazardous work activities (WEF, 2017). Designers face the formidable challenge of ensuring that the facilities they design meet all of the performance criteria while incorporating applicable safety and health considerations.

While complying with codes and regulations, the design engineer is also required to exercise sound judgment and use procedures that improve facility safety. These procedures include, but are not limited to, the following:

- Implementing good design practices that promote safety (e.g., using less hazardous or more dilute chemicals and facility layout to accommodate confined-space entry and lockout and tagout procedures)
- Carefully considering chemical system aspects
- Analyzing O&M tasks to develop layouts that will result in a simple and safe way to accomplish a potentially hazardous task
- Conducting operability reviews
- Selecting equipment to reduce noise
- Providing items to facilitate safety such as safety literature, training room, and safety signs
- Building physical models
- Incorporating good HVAC design
- Ensuring maintainability
- Creating effective evacuation plans
- Inspecting the constructed facility

Table 7.5 lists typical tasks that a designer may carry out during the various project phases to address health and safety issues.

The following is a detailed list of key safety and environmental programs that must be considered in design:

- Community right to know (ensures that the public is aware of hazardous chemicals that will be used on site)
- Clean Air Act (requires permits for air emissions from incinerators and other unit processes)

TABLE 7.5   Health and safety tasks for a designer (adapted from Water Environment Federation [2017]).

Project Phase	Task
Facility planning	Consider safety risks during treatment process selection and identify safeguards for the selected process
Preliminary design	Review existing accident reports, study code compliance, and conduct health and safety survey
Final design	Review plans and specifications for safety compliance
	Prepare construction documents, emergency response plan
	Incorporate a chapter on safety in the O&M manual
Construction	Ensure contractor is in compliance with the construction documents with respect to safety
	Conduct health and safety survey of completed project
Startup testing	Provide safety training

- Clean Air Act Hazardous Chemical Management
- Process safety management
- Laboratory chemical hygiene
- Lockout and tagout
- Confined-space entry

All facilities must be designed to meet Occupational Safety and Health Act requirements (29 CFR 1910 and 1926). Several industrial associations provide information for designers. These associations include the Chlorine Institute (Washington, D.C.), the Chemical Manufacturers Association (Arlington, Virginia), the National Fire Protection Association (Quincy, Massachusetts), and the National Safety Council (Itasca, Illinois). In addition, safety organizations such as the American Society of Safety Engineers (Fairfax, Virginia) and the American Industrial Hygiene Association (Des Plaines, Illinois) are active in many cities and states. Also important are required U.S. EPA and other codes, management and emergency response reporting, and detailed plan requirements for chemicals and other potentially hazardous materials.

6.13.1 Local Safety Codes

Many requirements for existing and proposed local fire codes are more detailed and stringent overall (or in specific areas) than Occupational Safety and Health Administration (OSHA) regulatory requirements (e.g., California OSHA). In some instances, local fire and other safety codes dictate layouts, construction materials, and safety equipment. Local fire departments typically have the authority to establish requirements for site layouts to allow access for emergency response equipment and firefighters; establish the number, location, and type of fire hydrants; and set other requirements. Care should be taken to communicate early and often with local code authorities and fully document those requirements.

The Uniform Fire Code establishes requirements for features to be included by the designer and, in some cases, requires specific analyses (e.g., seismic analyses) by the designer.

6.13.2 Americans with Disabilities Act Requirements: Facilities

Facilities under design or renovation must comply with federal Americans with Disabilities Act (ADA) regulations unless specific exemptions are allowed (for certain areas) by the body governing ADA requirements where the upgrade is taking place. The ADA covers accessibility issues, such as the use of handrails, ramps, parking facilities, and lavatories, and these must be accounted for during design.

6.14 Vulnerability Assessment

Wastewater and water treatment facilities in the United States are becoming increasingly concerned about their continued ability to provide public health and environmental protection. In addition, there is a need to ensure that the country's infrastructure is not used as a vehicle by terrorists to cause potential damage to the communities they serve. America's Water Infrastructure Act (AWIA), which became a law in 2018, requires each community water system serving more than 3,300 people to assess the risks to and resilience of its system to malevolent acts and natural hazards. Accordingly, utilities need an effective tool to assess vulnerabilities and to establish a risk-based methodology for making necessary changes. Several approaches are available for application to WRRFs, including the following:

- The vulnerability self-assessment tool (VSAT) designed by U.S. EPA to help water systems comply with AWIA

- Center for Chemical Process Safety's security vulnerability analysis methodology developed by the American Institute of Chemical Engineers
- Risk Analysis and Management for Critical Asset Protection (RAM-CAP®) Standard developed by the American Water Works Association (AWWA, 2010)

While either methodology may be used, VSAT is available free of charge to all public utilities on the U.S. EPA website. VSAT assesses vulnerability, prepares responses to extreme events, and restores normal business conditions thereafter. The key elements of VSAT include identifying and categorizing assets, determining criticality, assessing existing countermeasures, assigning vulnerability ratings, identifying risk level, identifying and estimating risk mitigation costs, and developing implementation plans. For a detailed discussion on these elements as well as tutorials on using VSAT, readers are encouraged to visit the U.S. EPA website.

These tools will help the owner and engineer evaluate the vulnerability and risk associated with a specific treatment facility or process. Often, specialists are hired separately by the owner to carry out insurance or risk assessment. Results of those assessments should also be incorporated in the design.

6.15 Site Conditions

Building on the information from the planning stage, during the preliminary design stage it is important to revisit and understand the history of the site and how it was used, the subsurface conditions, whether there are buried structures, and so on. If the site was used as a landfill, then it is important to do as much sampling and testing of soils and groundwater as possible. Soils and/or groundwater contamination will significantly affect the disposal options for both and can significantly increase project costs, permitting issues, and construction time. If there is a history of buildings on the site, then considerable effort should be made to locate buried structures that might result in change orders during the construction phase.

The more geotechnical work that is done at this stage, the less likelihood there is of change orders or design changes once the project begins. Money spent at this stage for geotechnical work will save money during the construction phase. The more that is understood about the site and site conditions—as early as possible but especially at this stage—the less chance there is to incur added and unknown costs during construction. At the very least, some borings and test wells should be inserted throughout the site to evaluate the type of foundations that need to be used (piles, spread footings, etc.). The test wells will give information on groundwater quantity and quality and help determine, with construction permit requirements, whether it can be discharged to the facility without prior treatment during construction dewatering.

Hazardous materials investigations and surveys will inform special project and demolition requirements, such as lead paint abatement, asbestos removal, and location of polychlorinated biphenyls or other hazardous materials that may need to be handled and disposed of during a modification or rehabilitation of existing facilities. It is important to document existing conditions so that the contractor includes special requirements in the pricing.

6.15.1 Building Requirements

It is important to decide what buildings will be required and whether old buildings can be upgraded for new uses. Building space considerations should include facility laboratory, men's and women's locker rooms, break rooms, kitchen areas, offices, conference and meeting rooms, storage (equipment and paperwork), indoor vehicle storage, maintenance shop, and so on. This may require a series of buildings throughout the site.

6.15.2 Utilities

Decisions need to be made on the fuel that will be used for heating buildings (gas, oil, or electric), heating processes (digester gas), and providing heat sources such as for a sludge-drying process. Furthermore, alternative or backup power sources will also be necessary. This requires two separate and distinct main power sources for the facility or a single power source and a backup emergency generator, which will run the entire process. Typically, the state regulatory agency will also have specific requirements for alternative power sources. It is important to discuss facility needs with the various utilities to ensure that the utility can meet the demands.

Other utilities include fiber-optics cable for data transmission and networking, intercom and telephone connections among buildings, and wireless communication systems. Furthermore, there should be a facility water distribution system to be used for various purposes instead of potable water, although a source of potable water must be identified and proper backflow prevention included as required.

7.0 QUALITY ASSURANCE AND QUALITY CONTROL

The functions of quality assurance (QA) and quality control (QC) for preliminary design are primarily focused on documenting the design calculations. Standard templates should be prepared for calculation documentation such as listing assumptions, formulae, criteria, and reference values. The calculation document should typically go through three phases: preparation, checking and review. As the design engineer completes a design task, the project manager or design manager should check the work to ensure that the design conforms to the project quality requirements. It is recommended that checks be performed after completion of each task. The last phase is the review of the design by a senior reviewer who is aware of the project goals, strategies, and planning. Typically, the reviewer would be someone with strong technical knowledge and experience with projects of similar or larger magnitude and complexity. The overall QA/QC process should result in a good-quality report of preliminary design that could also be perfected with protocols of good modeling practice applied during the preliminary design.

8.0 MANAGING COSTS

During the facility planning process, an initial estimate is established for the overall cost of the project. This cost includes both the engineering fee and construction costs. It also may include operational costs. As the design progresses, it is essential that both the engineer and owner not deviate from the basic decisions that formed the basis for that estimate. A change management process is needed to help both the engineer and owner monitor the project costs and provide early warning of cost increases. All too often, cost increases are the result of requests by the owner's representatives for changes or additions to the previously agreed-upon scope. An effective management technique is to develop a prioritized list of improvements, including "must do," "should do," and "nice to do, if there are adequate resources." This prioritized list, which can be developed by the stakeholders, will then serve as a means of ensuring that the most important aspects of the project are addressed if resources become an issue.

A second major cause of cost increases is a facility plan that is a poorly or vaguely defined. A third cause is unforeseen conditions. Facility upgrade projects are particularly prone to higher costs because of poor as-built drawings of past construction, deteriorated structures that were not anticipated for replacement in the upgraded design, a need to work within the internals of the facilities and equipment while maintaining operations, and poor records of underground conditions. Extra design effort is required to identify and plan for as many of these circumstances as possible upfront to save time and money, prevent change orders, and minimize potential for construction claims.

9.0   REFERENCES

American Water Works Association. (2010). *Risk Analysis and Management for Critical Asset Protection (RAMCAP®) Standard for risk and resilience management of water and wastewater systems using the ASME-ITI RAMCAP Plus® methodology* (1st ed., ANSI/AWWA J100-10 [R13]). American Water Works Association.

Andraka, D. (2020). Reliability analysis of activated sludge process by means of biokinetic modelling and simulation results. *Water, 12*(1), 291.

Corbitt, R. A. (1998). *Standard handbook of environmental engineering* (2nd ed.). McGraw-Hill.

Metcalf & Eddy, Inc. (2014). *Wastewater engineering: Treatment and reuse*, (5th ed.). McGraw-Hill.

Qasim, S. R. (1999). *Wastewater treatment plants: Planning, design, and operation* (2nd ed.). Technomic.

U.S. Army Corps of Engineers. (2008). *Engineering Guidance Design Manual* (CEHND 1110-1-1). https://www.hnc.usace.army.mil/Portals/65/docs/Directorates/ED/DesingManual/Locked%20Mar%203%202008%20Design%20Manual.pdf

U.S. Environmental Protection Agency. (1982). *Evaluation and documentation of mechanical reliability of conventional wastewater treatment plant components*; (EPA-600/2-82-044). U.S. EPA.

Wastewater Committee of the Great Lakes-Upper Mississippi River. (2014). *Recommended standards for wastewater facilities*. Health Research, Inc.

Water Environment Federation (2014). *Wastewater treatment process modeling* (2nd ed., Manual of Practice No. 31). Water Environment Federation.

Water Environment Federation. (2017). *Design of water resource recovery facilities* (6th ed., Manual of Practice No. 8). Water Environment Federation.

10.0   SUGGESTED READINGS

Daigger, G. T., & Buttz, J. A. (1998) *Upgrading wastewater treatment plants*. Technomic.

Environment Canada. (2006). *Guidance manual for sewage treatment plant process audits*. Environmental Technology Advancement Directorate.

8

Final Design Phase

John Scheri, PE, BCEE & Amber Schrum

1.0 INTRODUCTION

The final design phase is one of several critical phases in delivering a water resource recovery facility (WRRF) project. It requires the most attention to detail and often takes the longest to complete. In general, the final design phase includes advancement of the preliminary design to 60%, 90%, and 100% completion. Design progress reviews and checking are typically performed after 60% and 90% completion. It is where many mistakes can be made. Therefore, adequate time and resources are critical to the success of this phase.

2.0 MANAGEMENT OF THE FINAL DESIGN PHASE

Participation and cooperation are required at all levels to ensure a successful project, which is delivered on time, within budget, and to the owner's satisfaction. The participants typically involved in the final design phase are described in Table 8.1.

Figure 8.1 below illustrates the relationships between various project stakeholders.

2.1 Project Leadership

The owner will provide leadership by setting the goals and overall schedule for the project. The owner is responsible for obtaining the funding, retaining

TABLE 8.1 Participants during final design phase.

Role	Considerations
Owner (city, utility, private concern, or other) and operators	Express the need for the project and drives the project forward. They determine an inadequacy in the current state of their facility because of a regulatory requirement or operational need, and have decided to complete the project. The owner or operator(s) may or may not explicitly share their opinions or biases with the engineer. However, these must be addressed in the design for the project to be successful. Failure to incorporate these may lead to dissatisfaction and resentment from facility operators, which could eventually lead to project failure as the operators, either consciously or unconsciously, make the project fail. Other owner staff may coordinate to achieve goals related to sustainability or energy.
Engineer	The WRRF designer who provides the technical expertise and the staff/time resources to complete the design in a timely fashion. It is the engineer's responsibility to communicate with the owner or operators to effectively determine their preferences.
Regulatory agencies	May indicate the need for operations improvements or facility upgrades based on their observations and analysis. They provide treatment objectives, set boundaries (time limits, etc.) and, in some cases, provide funding. Some agencies can set unrealistic or impractical goals and objectives. It is far better to reason with these agencies than to litigate. Examples: U.S. EPA, state environmental regulatory agency, U.S. Army Corps of Engineers, Department of Fish and Wildlife.
Local agencies	May inspect for compliance with building codes, issue certificates of occupancy, or provide a source of outside quality control. Examples: city/county engineers, planning departments, building departments, planning or zoning boards, historic design review commissions, wetland conservation committees, soil erosion and sediment control agencies, stormwater agencies, floodplain managers, utilities, emergency services.
Funding agencies and finance professionals	Sources with money for grants or low-interest loans. They may influence process selection, especially if they believe a process is unproven. This category may also include a financial advisor and bond council who determines the funding type and amount that can reasonably be used for a project. See Chapter 7 for a complete discussion of funding and financing.

(continued)

TABLE 8.1 (*Continued*)

Role	Considerations
Owner's attorneys or legal specialists	Advise on legal implications of certain actions, procedures, or practices (e.g., procurement or bidding practices). They provide legal opinions to funding agencies and represent the owner on claims, if any. Legal advice should also be provided if any alternative delivery method or nonstandard bidding process will be used.
Public relations	Ensures that the public understands that the optimum technology is being used, and is safe, effective, reliable, and economical. Communicating the need for, location of, and specific processes used in a WRRF are essential to community buy-in. A consistent message delivered competently, confidently, and by trustworthy individuals is critical to success. Those responsible for public relations may be internal employees or special consultants. With the rise of social media, some utilities have used interactive websites to distribute information and notify the public regarding construction progress and potential impacts, such as road closures.
Neighbors	Want an invisible and odorless facility, or at least a facility that minimally affects their lives and property. The owner will need to provide guidance to the engineer so that appropriate aesthetic amenities are included in the design to mitigate impacts to neighboring properties.
Equipment manufacturers/sales representatives	Want their equipment to be included in the design. They provide information to the owner and operator regarding equipment, and provide technical data needed by the engineer to prepare the design.
Academia	Researching innovative processes or would like to support bench scale-testing, on-site pilot studies, or sampling of emerging constituents.

the professionals, and accepting and operating the facilities when they are complete. During the actual execution of the project, there are also technical leadership roles often delegated by the owner to the engineering, financial, and legal professionals. The engineer is responsible for the detailed design and often leads the preparation of cost opinions and contract documents that the owner and its project delivery team need.

2.2 Regulatory Agencies and Permitting

Open communication with regulatory agencies can speed the design process. Regulatory agencies define the objectives (e.g., effluent limits, time limits);

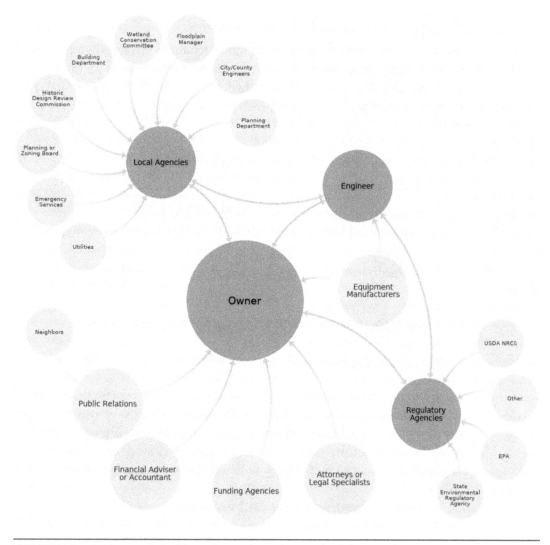

FIGURE 8.1   Project stakeholder relationships during final design phase.

may define allowable treatment processes; regulate the bidding process; or, in some cases, supplement funding. Depending on the jurisdiction and necessary permits for the project, permitting may commence during the preliminary design phase or in conjunction with the final design phase. Permits can include wetland and floodplain disturbance, cultural heritage reviews, treatment changes, treatment works construction permits, air emission permits, stormwater permits, and local building and traffic permits. The earlier that comments are received from the regulatory agency and incorporated into the design, the better that large design changes and the resulting budget impacts can be limited.

2.3 Interaction with Operational Staff

It is necessary to obtain operator and maintenance staff buy-in and ownership of the facility. Operators and maintenance staff should be involved in all design review workshops, described in further detail in Section 7.1 of this chapter. Key areas for operational staff interaction include, but are not limited to:

- Maintenance of facility operations during construction
- Safety
- Equipment accessibility for ease of operation and for maintenance
- Degree of automation, supervisory control and data acquisition (SCADA)
- Standardization of equipment based on local sources of supply

3.0 PROJECT CONTROLS ADVANCEMENT

3.1 Project Scope of Work

The project scope of work is a detailed description of project goals, objectives, and deliverables. It is also a descriptive, complete list of what the engineer is to provide for design and project delivery services. A complete scope, mutually agreed on by the owner and engineer, is a critical aspect of the final design phase and will greatly reduce misunderstandings and surprises.

3.2 Codes, Standards, and Best Practices

Codes and standards identified by the designer during the preliminary design must be carried through the final design of the project. See Chapter 7 for a complete discussion of codes and regulations. These may include:

- Permit criteria (e.g., allowable effluent limits)
- Local building codes (e.g., building, electrical, plumbing, fire protection)
- National Fire Protect Act (NFPA); Occupational Safety and Health Act (OSHA) and the Americans with Disabilities Act (ADA) for public areas (e.g., administration buildings)

The designer must also comply with, and incorporate, owner standards throughout the project design. Design guidelines, such as Water Environment Federation Manuals of Practice, Ten States Standards, and *TR-16 Guides*

(New England Interstate Water Pollution Control Commission, 2016) are common references for design, and are sometimes adopted by the regulatory agencies having jurisdiction.

3.3   Schedule for Milestones and Submittals

Submittals can include 60% and 90% detailed design, issue for regulatory review, or issue for construction. Milestone examples are shown in Figure 8.2. It is advisable to pair these submittals with an in-person design coordination meeting with the owner and outside stakeholders if required. Public meetings may be necessary.

Bid phase services for traditional project delivery typically include bid advertisement, pre-bid meeting, addenda preparation, bid opening, contract execution, and conformed document preparation. Design phase submittals should not be confused with construction submittals, which the contractor uses during construction to seek approval and provide coordination for materials and methods to be used. See Chapter 9 for a discussion of construction submittals.

3.4   Budget Control

All parties must have a clear understanding of project budgets. During the design phase, budgets can be broken down into two components: engineer budget and construction budget. It is the engineer's responsibility to stay within the agreed-upon or contracted budget for the given design scope. It is the responsibility of the owner to ensure that the scope is met and to ensure that the owner's staff does not make unwarranted changes or demands (out-of-scope work) that will increase engineering costs. The owner should recognize that changes in scope or design can affect schedule, engineering

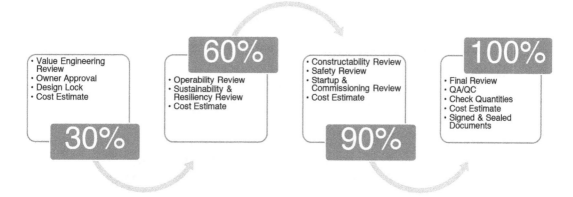

FIGURE 8.2   Project design schedule milestones.

costs, and construction costs. The engineer should inform the owner of the effects of design changes on scheduling and costs. Because of the potential for costly last-minute design changes, consider maintaining a scope change register to aid in discussing out-of-scope work, and fee and schedule adjustments. It is typical for 20% of the design budget to be exhausted preparing the final 10% of the design. This is because of the extensive coordination and checking that must be done to complete the final design, without appreciable change to the design plans and specifications.

During the preliminary and final design phase, the construction budget can be treated as a target, which the project construction bids are intended to match or fall below. The engineer must manage "scope creep" by avoiding incorporation of additional processes, equipment, or other features that will cause the construction budget to be exceeded. The budget must be met, and owner expectations must be met or exceeded. If it is determined that the project will not meet or exceed owner expectations, it may be necessary to modify the project scope and negotiate a suitable design contract amendment to provide more resources to deliver a quality project.

4.0   CONSTRUCTION DOCUMENT PREPARATION

Construction documents (drawings and specifications) are the legal description of work to be completed by the contractor or other parties, and dictate the expected level of performance. Simple and easy-to-read documents that are as complete as possible will help bidders to not overlook work, which will limit the potential for change orders as omissions are minimized.

4.1   Technology Plan

There are many computer programs used by engineers during the design preparation for WRRFs. Depending on project magnitude and complexity, there may be many software programs used across multiple disciplines, and the project design team may include staff in multiple locations or across firms. It is critical that the design team follow a project-specific technology plan through the preliminary and final design phases. The technology plan defines the software to be used and the conventions for each application so that a common data environment exists for the project. It is essential to define what content will be created in two-dimensional or three-dimensional computer-aided design versus three-dimensional building information modeling (BIM) models, what software packages will be used, and which version. The coordination of software versions between disciplines is now an essential part of project management, as many programs prevent back-saving to previous versions and files saved in different versions can create linking errors. Some software enables coordination review between disciplines, and this should be considered when choosing software.

Of equal importance in the technology plan is where data and working models will be saved, and their naming conventions. This attention to detail makes locating files easier, simplifies making design changes, and facilitates the preparation of final documents after construction is completed, such as standard operating procedures (SOPs), operations and maintenance (O&M) manuals, and record plans.

4.2   Hard Copy Versus Digital Deliverables

Whether the client has requested hard copy or digital deliverables will determine the level of complexity and detail used in modeling the project for contract document preparation. In the case of three-dimensional or BIM models, contract drawings are prepared from sections and plan views of the three-dimensional model, as shown in Figure 8.3. However, some clients may require turnover of the model for additional analysis of building performance, life-cycle assessments, and creation of bills of material and cost estimates. The differences in modeling effort mean that this must be defined in the engineer's scope of work. The typical level of development (LOD) for traditional WRRF design–bid–build projects is LOD 350, as defined by the National Institute of Building Sciences' *National BIM Standards-US* (NIBS, 2015). Furthermore, the owner should clearly define the raw, editable format of the deliverable that the engineer is to provide upon completion of the design.

Bid documents can be produced in electronic PDF format as a supplement or alternative to hard copies of drawings and specifications. The hard copies should be considered the official documents in case of conflicts. BIM and/or AutoCAD files of drawings can be provided to the contractor to assist in preparing shop drawings, but this should be specified in the engineer's scope of work. If editable electronic files are to be provided to the contractor, the owner should consult with their legal counsel to prepare a mutual release of liability for the contractor's use of these documents.

4.3   Drawings

Simplicity is important, as cluttered drawings inevitably cause oversight. In today's world of computer-aided and computer model-based design and drafting, there is little justification for packing the maximum amount of information onto the minimum number of sheets, and goals of reducing sheet count are self-defeating in the long term. In general, the more that can be reasonably shown on the drawings, the better, because they are the most

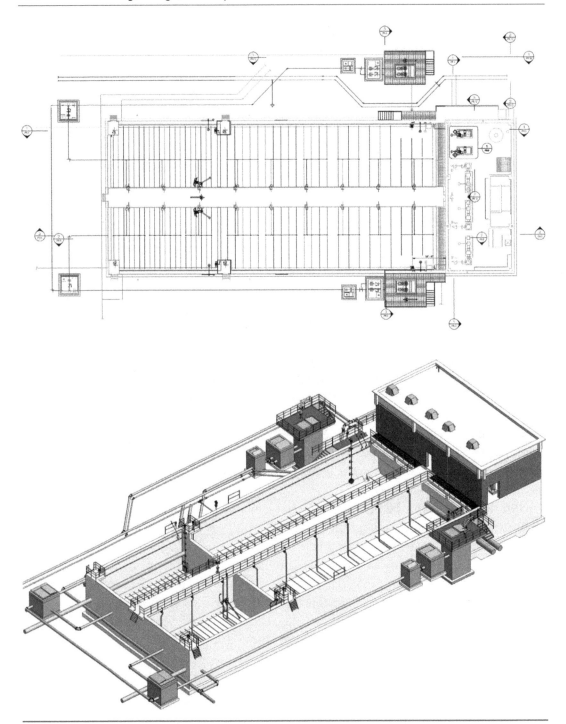

FIGURE 8.3   Sample two-dimensional drawing versus three-dimensional rendering (courtesy of Hanover Sewerage Authority, Whippany, New Jersey).

used documents in the field. It is generally advantageous (and, in many cases, required) to provide separate drawings for each engineering or architectural discipline, even in locations where engineering or architectural work must share a common area. This enables bidders and sub-bidders to focus on their specialty, in addition to providing clarity. The set of drawings should include an index listing the individual sheets and the title of each sheet. Large projects may include multiple volumes of drawings separated by construction disciplines or specific work packages. Drawings grouped by discipline may make it difficult to locate plans for a specific area without referring repeatedly to an index or scanning through multiple drawings. In the case of large projects, a prefix number can be used to define an area or process at the facility. Therefore, all processes or areas of the facility would have a specific group of drawings associated with them, and a key plan provided to denote which area is assigned to each prefix number. Design drawing sets for WRRF projects are typically organized by discipline, as illustrated in Table 8.2.

Separating drawing sets should be discouraged because this can inhibit coordination between construction disciplines if the entire package is not available. It should be noted that some states require separate contracts for general construction; heating, ventilation, and air conditioning (HVAC); electrical; and plumbing work, which may require separating contact documents by discipline.

Depending on project complexity, the drawing number may also be used to denote whether it is a plan, section, or detail sheet. Many companies prefer to utilize the *National CAD Standard* (NIBS, 2014) or British Standards Organization (BSO, 2019) drawing number conventions modified to their specific needs.

The development of drawings is a time-consuming process, wherein coordination among disciplines and specifications is critical. Detailed discussion of drawing development is beyond the scope of this chapter. Drawings should include detailed plan views, enough sections so that the contractor can adequately bid and construct the project, special details, and standard details. These should be arranged and cross-referenced using conventions that are easily understood.

The preparation of lists and schedules is also an important aspect of the final design phase, which can be time-consuming and become the source of errors. The desire for lists on the part of designers, bidders, and equipment providers must be balanced by the actual need to have such lists, especially when bidders should be doing their own takeoffs during the bidding process and determining the orientation, location, size, and details of a piece of equipment based on the drawings.

Finally, the engineer of record or responsible engineer should seal the drawings. Often, drawings will be sealed by two or more engineers and/or

TABLE 8.2 Drawing discipline abbreviations.

Discipline	US National CAD Standard Abbreviation	BS EN ISO (19650-2:2018) Abbreviation	Other Abbreviation
General	G	Z	
Site/Civil	C	C	
Traffic control	—	—	T
Landscaping	L	L	
Demolition	—	—	D, DM
Process: mechanical	D	P	PR, M
Structural	S	S	ST
Architectural	A	A	AR
Electrical	E	E	EL
Instrumentation and controls	—	—	I, PI
Building: mechanical or HVAC	M	H	
Plumbing	P	—	M, PL
Fire protection/life safety	F	—	FP

Source: NIBS (2014) and BSI (2019).

an architect, depending on the engineering disciplines required to produce the drawings.

4.4   Specifications

The engineer must produce specifications that serve as a written guide for equipment procurement, construction materials, and construction methods. These documents are typically bound and are part of the construction contract. The project specifications may include standard specifications with specialty sections for facility-specific or designed equipment. The project specifications may be a mix of owner's specifications, engineer's specifications, American Institute of Architects (AIA) specifications, and specifications from other sources. The use of standard specifications with standard formats (e.g., Construction Specifications Institute [CSI] or AIA specifications) will minimize confusion by prospective bidders. Specification format and appearance should be consistent, where possible, throughout the entire set of specification documents.

4.4.1   Front End

Front-end specifications include sections such as the invitation to bid, instructions to bidders, bid forms, sample contract performance and payment bonds, the contract, basis of contract award, risk allocation, definition of changed conditions, definition of substantial completion, and general and supplementary conditions. Wage rates, if applicable, should also be included here. Some of these may be owner-provided, but their assembly into the contract documents may be handled by the engineer.

The "front end" (as the contractual and general project requirements are generally known) may be a mix of standard documents, industry documents (e.g., Engineers Joint Contract Documents Committee [EJCDC]), or the owner's preferred documents; in any case, they must be refined for individual projects. It is strongly advised that the owner's legal counsel review these front end documents during the final design phase when the design nears 100% completion.

Industry-accepted standard construction contracts are available for WRRF construction. Several institutes have published contracts and associated contract management forms, including AIA, ConsensusDocs, Design–Build Institute of America, and EJCDC.

Standard contracts are preferable to an owner's custom in-house contracts because they are "tried and true"; well-vetted by owners, contractors, and engineers; regularly updated to reflect current legal standards; and cover all aspects of WRRF construction (whether small or large). The use of a standard contract reduces legal review periods and reduces risk allocation made by a contractor when pricing the work.

A standard contract (aka general conditions) is tailored to the specific WRRF project by preparing a supplemental document that amends specific contract language or adds provisions unique to the project. Otherwise the standard contract language remains unchanged.

State and/or local public contract bidding laws have specific requirements that must be incorporated into the specifications. It is essential that the bidding forms and instructions are clear to prospective bidders so that the bids are submitted correctly. Incorrect bid submissions can contain non-waivable fatal defects, which require the owner to re-advertise and rebid the project.

4.4.2   General Project Requirements

The front end of the specifications should also include general project requirements, such as measurement and payment, payment procedures, project coordination, progress schedules, construction progress photograph requirements, submittal procedures, hazardous locations, health and safety requirements, regulatory requirements, reference standards, temporary construction facilities or utilities needed, protection of stored materials, maintenance of facility operations during construction, equipment installation, cleaning and site maintenance, preliminary and final field tests, equipment startup and training, O&M manual requirements, and project closeout. The specifications should call for a training plan or program to address training for equipment and systems; overall process training; and associated instructor and material submittal requirements, scheduling, and documentation. Requirements should be stipulated for the training duration, classroom location and environment (e.g., no noise, painting, or other disruptive construction activities). Training final documentation, attendance records, training hours accreditation for license renewal, and training videos should be specified. An additional discussion of commissioning specific specification requirements can be found in Chapter 10.

The engineer will specify the required submittals in the contract specifications. It is the contractor's responsibility to provide all submittals and respond to the engineer and owner's comments as applicable.

The specifications should define who will be responsible for obtaining regulatory approvals and construction permits. It is typical for the owner or engineer to obtain major regulatory approvals for the construction of WRRFs; however, the construction contractor will typically apply for and obtain building permits. It should be noted that dewatering permits may be required in advance by the owner or engineer if extensive dewatering is required. However, it is also common for the construction contractor to be responsible for this permit. The extent and responsibility for dewatering should be reviewed carefully by the engineer so that responsibility can be appropriately assigned and specified.

4.4.3   Other Technical Data

Supplemental information to assist contractors in preparing bids, such as geotechnical reports, soil test boring logs, and hazardous materials analysis reports, can be included as appendices to the contract documents. The contract documents should include a disclaimer that the subsurface investigation reports are for information only and are not part of the contract.

4.4.4   Technical Specifications

The final design includes the preparation of detailed specifications for materials, products, and equipment to be used for project construction. WRRF construction projects are typically bid as a lump-sum price, so specifications are traditionally organized by building trade or design discipline. Other sectors (e.g., transportation) are often bid as unit price contracts and have itemized specifications. In either model, specifications are interrelated, and special attention should be given to cross-references.

Specifications for WRRF projects can encompass most design disciplines and are specialized because of a strong emphasis on the equipment to be provided. Technical specifications can be written as prescriptive or as performance based. Performance-based specifications may be supplemented by a prequalification phase for manufacturers. See Section 5.4.1 in this chapter for more details on equipment specification.

The designation of each specification section has evolved from 5 divisions to 16 divisions, and most recently, to the current industry standard of 50 divisions. The following table has been prepared from a variety of sources (CSI's *MasterFormat* [2018]; AIA's *MasterSpec* [n.d.]) to generally correlate the transition from the 16-division specification to the 50-division specification.

4.5   Project Reviews

Special attention must be given to coordination among disciplines, which ensures that electrical drawings have picked up all electrical loads shown on other drawings, weights and dimensions of HVAC equipment are accounted for by structural drawings, conflicts between disciplines are checked, and so on. This coordination can also be completed effectively during design workshops paired with BIM models attended by a designer and a modeler from each discipline. These workshops facilitate real-time alterations, allowing disciplines to collaborate in identifying potential conflicts and workable solutions.

Drawing coordination is ultimately the engineer's responsibility, but all parties should evaluate drawings and specifications during the final design phase and look for, as a representative list, the following:

- Electrical or control panels or elements that were changed or moved
- Equipment and piping that was moved, renamed, or assigned some other use
- Motor sizes or speeds that have been changed
- Control valves that were moved or renumbered
- Consistent entry points for pipes and conduits
- HVAC louvers that were relocated

Many of these items can be changed by one discipline at the direction of the owner, and any change late in the design phase opens up greater possibilities for conflicts in the contract documents. Linked BIM models can also facilitate coordination reviews at this stage.

TABLE 8.3   Comparison of 16-division to 50-division specifications.

16 Division	50 Division
00 – Front End	00 – Procurement and Contracting Requirements
01 – General Requirements	01 – General Requirements
02 – Site Construction	02 – Existing Conditions
	31 – Earthwork
	32 – Exterior Improvements
	33 – Utilities
	35 – Waterway and Marine Construction
03 – Concrete	03 – Concrete
04 – Masonry	04 – Masonry
05 – Metals	05 – Metals
06 – Wood and Plastics	06 – Wood, Plastics, and Composites
07 – Thermal and Moisture Protection	07 – Thermal and Moisture Protection
08 – Doors and Windows	08 – Openings
	28 – Electronic Safety and Security
09 – Finishes	09 – Finishes
10 – Specialties	10 – Specialties
11 – Equipment	11 – Equipment
	41 – Material Processing and Handling Equipment
	42 – Process Heating, Cooling and Drying Equipment
	43 – Process Gas and Liquid Handling, Purification, and Storage Equipment
	44 – Pollution and Waste Control Equipment
	45 – Industry-Specific Manufacturing Equipment
	46 – Water and Wastewater Equipment
12 – Furnishings	12 – Furnishings
13 – Special Construction	13 – Special Construction
	28 – Electronic Safety and Security
	40 – Process Interconnections
14 – Conveying Systems	14 – Conveying Equipment
	34 – Transportation

(*continued*)

TABLE 8.3 Comparison of 16-division to 50-division specifications. (*Continued*)

16 Division	50 Division
15 – Mechanical	21 – Fire Suppression
	22 – Plumbing
	23 – Heating, Ventilation, and Air Conditioning
	25 – Integrated Automation
	40 – Process Interconnections
16 – Electrical	26 – Electrical
	27 – Communications
	48 – Electrical Power Generation

5.0 FINALIZING PROCESS SELECTION AND DESIGN

Finalizing process and equipment selection is often easier said than done. A variety of process advantages and disadvantages, equipment features, and competing interests must all be addressed. In general, the final process selection and design should be a follow-through of the facility plan and 30% preliminary design. The facility plan generally selects the unit processes that will be used to achieve treatment goals. Too much deviation from the facility plan may result in difficulties during review by agencies that had approved the facility plan. The facility plan may have to be modified to meet the design, or the design may have to be modified—possibly at great cost—to meet the facility plan. The final process selection and design should develop to completion the concepts outlined during the preliminary design phase. Final design can be condensed into the steps described in the following sections.

5.1 Process and Instrumentation Diagrams

Process and instrumentation diagrams (P&IDs) depict unit processes in sequential order and their interrelationships with each other. Process flow diagrams start as simple drawings and become increasingly complex until they develop into P&IDs. The P&IDs show unit processes with all of their interconnections, monitoring, and control components, and rudimentary control logic. P&IDs can form the basis for planning and designing control concepts to be incorporated into automatic control and monitoring systems. These drawings are initiated during preliminary design and completed during final design.

5.2   Hydraulic Profile

The initial hydraulic profile is typically developed during preliminary design and finalized during the early part of final design. Hydraulic profiles should be developed for average and peak flows, and all internal recycle streams must be considered. Extra fall should be allowed at critical locations. At the final design stage, it is important to check all facility data to verify that there have been no changes in influent flow and characteristics since the planning phase. It is especially difficult to develop a hydraulic profile in facility upgrades where the gradient, pipe and structure capacities, and bottlenecks from previous designs must be adapted for future capacity. Detailed closed-pipe, open-channel, and free-fall (i.e., weir) calculations should be provided and double-checked by the engineer. Computer modeling and simulation may be warranted.

In the hydraulic profile, consideration should be given to future treatment units that may be necessary in future facility improvements. For example, there may be a pending regulation for tertiary filters, and it may be wise to include sufficient fall in the hydraulic profile to accommodate these future facilities.

5.3   Basis of Design Report

During the final design phase, the basis of design report (BODR) should be brought to completion to serve as the guide for developing final design deliverables. The BODR should finalize outstanding characterization data, address remaining interpretations of applicable building codes and standards, confirm materials selection for all major equipment and process systems, and incorporate project specification requirements dictated by regulations and/or permit conditions.

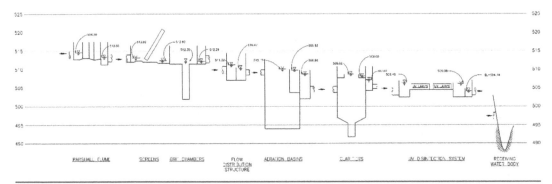

FIGURE 8.4   Example hydraulic profile (elevations shown in feet above mean sea level; water surface elevations shown for 5 mgd flow).

5.4   Equipment Selection

Major equipment is selected during preliminary design and finalized during final design. In addition, equipment not already selected will be chosen during final design. The selection of some equipment might be the result of side-by-side testing or on-site pilot testing. Owner participation in the process is critical, as the owner will have opinions about equipment type, manufacturers, and sales representatives.

Criteria that should be considered during equipment selection include service and parts availability, compatibility with existing systems and equipment, and operator knowledge and experience. The chosen equipment may become outdated by the time the project reaches the construction stage, particularly process instrumentation. The construction specifications should be worded so that substitutions with an updated model or version can be made later. Because of accelerating technological advancements, design engineers must make a concerted effort to evaluate new technology by reading industry publications, participating in professional organization events, and maintaining relationships with manufacturers' representatives. References of interest for equipment design issues include *Wastewater Engineering: Treatment and Resource Recovery* (Metcalf & Eddy, 2014) and *Design of Water Resource Recovery Facilities* (WEF, 2017).

The engineer and owner must evaluate together the need for standard versus specialized equipment. Equipment that is "off the shelf" will be less expensive, easier to obtain parts for, and less likely to need highly specialized or trained maintenance personnel. When it makes sense, equipment that is identical to existing equipment should be selected to reduce spare parts inventory and simplify maintenance training. Where practical, selection of equipment that is unique, proprietary, or uncommon should be avoided unless it is sufficiently innovative to improve efficiency. This type of selection needs to be preceded by careful engineering analysis.

5.4.1 Equipment Specifications

The equipment specifications can be performance-based or prescriptive. Equipment specifications typically have three parts. Part 1 includes general requirements that describe administrative matters, such as related specifications, codes, standards, submittals, warranties, and quality assurance requirements. Part 2 includes product requirements, such as fabrication material, manufacturing tolerances, certified test curves, factory testing requirements, spare parts, lubricants, special tools, voltages, and the manufacturer's quality assurance and quality control (QA/QC) requirements. This is where the equipment manufacturer(s), product(s), and model number(s) used as the basis of design would be identified. Part 3 includes execution requirements and generally describes the work to be performed at the project site. This part describes delivery, storage, field installation, startup and testing requirements, quality control, and the manufacturer's services such as training. See the related discussion in Chapter 9.

5.4.2 Preselecting or Purchasing Equipment

In the case of equipment with long lead times, the equipment may be preselected or purchased to speed up the construction schedule or provide a guaranteed level of performance that cannot be easily obtained through a competitive bidding process. These acquisition processes also provide a means for the use of proprietary equipment. There are often state or federal requirements that must be followed if preselection is chosen. The owner and the engineer should discuss the merits of prepurchasing or preselecting equipment so that the appropriate information is included in the design.

5.5 Control Concepts

The control concept developed during preliminary design should be expanded in detail and finalized. Control concepts can range from very simple (manual operation) to very complex (fully automated). The level of control complexity will depend on facility size, equipment types, available staffing, and budget. Furthermore, the level of control complexity will depend on the owner's comfort level or experience with the control concepts and equipment proposed. Some facility personnel who have not had a lot of automated control experience may be uncomfortable with the new technology, and failure to account for this in design and postconstruction services (e.g., operating manuals and training) may lead to failure.

5.5.1 Normal and Alternative Operations

During final design, more details will be developed to describe how the facilities will operate during different conditions. For instance, the facilities may be designed to have dry- and wet-weather operating protocols. The different modes of operation can be automated or require operator interaction to be implemented. Modern facilities are typically designed to operate automatically based on time intervals or instrumentation inputs. Most equipment is provided with a hand/off/auto operator interface switch on a manufacturer-supplied or custom control panel. However, the equipment control panel may not be located next to the equipment, so it is advisable—and may be required—to have a local panel and emergency stop switch for operator safety.

Normal operations include the processes and equipment that are typically in service. Alternate modes of operation could include guidance for operations when a unit is out of service for maintenance or instructions on how to change the treatment process from one mode to another. For example, biological activated sludge facilities can have design features that allow for operation as traditional plug flow reactors, contact stabilization, or step feed.

The final design will typically include a functional description of equipment and instrumentation for normal and alternate modes of operation. The SOPs are then further defined in the facility O&M manual.

5.5.2 Emergency and Hazardous Operations

The final design should consider how the facility will be operated during emergency or hazardous conditions. These conditions can occur naturally (e.g., extended power outages, extreme weather events, earthquakes, hurricanes) or as a result of other factors (e.g., sabotage, terrorist attack, fire). The criticality of various design components will need to be assessed by the design engineer, operator, and owner to determine which facilities are required for emergency and hazardous operations. Features such as standby power, equipment operator switches, or manual controls will need to be designed to provide the means to continue operations.

5.6 Instrumentation, Control, and Automation; Supervisory Control and Data Acquisition; and Data Management

Modern control systems for WRRFs have moved away from control panels and wall-diagram displays to computer-based systems. Computer systems used for monitoring and control are called SCADA systems. New SCADA systems are often equipped with desktop consoles/computers and large-screen graphic displays. These systems are generally cost-effective to incorporate as part of a WRRF upgrade project and have the advantage of providing operators with user-friendly interfaces, improved data storage capability, and the information needed to effectively troubleshoot operational issues. See Chapter 7 for additional discussion on the benefits of operational data collection and uses.

Design of the SCADA system is completed during the final design phase. The basic control logic of the SCADA system is derived from the P&IDs, and the facility controls and data output are represented graphically on a computer monitor. The appearance and complexity of the graphical output should be developed in close coordination between the engineer and the owner. The design engineer and the facility operators should collaborate on determining a list of instruments that are needed to control the wastewater treatment process. There is growing interest for in-process water quality instrumentation to further automate the process. For example, measuring key process parameters can enable the adjustment of process aeration rates to save energy costs. Typical parameters measured at a WRRF include the following:

- Wet well levels
- Influent, air, return activated sludge, and waste activated sludge flows
- Aeration tank parameters such as dissolved oxygen, nitrogen, or oxidation–reduction potential
- Clarifier sludge levels, suspended solids concentrations
- Effluent chlorine residuals or ultraviolet transmittance
- Digester level, temperature, and gas pressure

Procurement of computer and control equipment differs from process equipment and requires careful attention to compatibility and upgradability; it relies heavily on sole-source procurement to match existing equipment. Data record increments (e.g., 5-minute, hourly, daily) should be determined during design, as adequate data storage must be provided. Data storage will also be affected by how long the data will be stored, and whether they will be saved forever, exported to another system, or overwritten after a period of time. The final data use will inform these decisions. It is common for these data to be used in regulatory reporting, planning for future upgrades or permitting, troubleshooting process upsets, and optimizing efficiency at the facility, such as potential cost savings from improved chemical or electricity usage.

6.0   FACILITY DESIGN

6.1   Site Layout and Considerations

The site layout at a WRRF should consider, as a minimum, the following items or issues, and should be developed with input from facility staff:

- The location of treatment processes and other structures must consider codes or rules for isolation from the general public and setback from public right-of-way and private property. Local zoning laws may address these requirements.
- Building/tankage layout and arrangement, and utility routing must be considered.
- The facility must comply with floodplain limits and required protection.
- Storm drainage systems must comply with state or local regulations and best management practices. Access for inspection, maintenance,

and compliance sampling at stormwater outfalls should be provided as required for facility regulatory compliance. Drainage from areas that may become contaminated by chemical or sludge spills should be directed to the head of the facility, accounting for spill containment as necessary.

- Facility driveways should provide access to all areas. Pavement should be designed for frequent and heavyweight traffic. Access is important for emergency response; chemical deliveries; residuals hauling; and equipment maintenance, rehabilitation, and replacement, including crane access for lifting out major equipment.

- Sidewalks should be located to allow visitors to readily identify safe pathways to a reception area or office and to provide connectivity throughout the facility between processes so that operators can safely make their rounds.

- Landscaping should be aesthetic—functional (e.g., prevent erosion and provide shade) and, most importantly, easy to maintain. See Chapter 7 for a complete discussion.

- Future upgrades, including adding new facilities or treatment processes, should be considered.

6.1.1   Environmental and Geotechnical Considerations

Climatic considerations are essential for successful treatment operations, such as whether process units need to be heat traced and insulated to handle low temperatures (to prevent freezing and facilitate nitrification) or enclosed to protect from snow accumulation or leaves falling in open tanks.

Geotechnical investigations will inform the layout of the facility and design of structural elements based on the depth of rock, its bearing capacity, the level of groundwater, and seismic conditions. See Chapter 7 for additional discussion on geotechnical investigations. Hazardous materials investigations and surveys will inform special project and demolition requirements, such as lead paint abatement, asbestos removal, and location of polychlorinated biphenyls or other hazardous materials that may need to be handled and disposed of during a modification or rehabilitation of existing facilities. It is important to document existing conditions so that the contractor includes special requirements in the pricing.

6.1.2   Site and Digital Security Requirements

It is important to have a secure fence, with card-key entry or security lock, around the WRRF to deter access to open tanks and process facilities that are generally unsafe for the public, and to protect the facility from vandals and other outsiders. The fences should be high enough so that no one can jump over them. Even small facilities need to control entry to the treatment facility. Trespassers could throw things into tanks that may damage equipment or interrupt treatment. Some treatment facilities have dangerous chemicals on-site, such as gaseous chlorine, sodium hydroxide, or methanol. These chemicals must be secured in such a way that no unauthorized personnel can access them. All buildings housing critical equipment or records should have coded or card-key entries and hazardous warning signs, such as confined-space entry and other signs or markings required by codes and regulations.

The facility should be well lit and have security cameras at various locations, especially areas where property damage can affect the process or where public health could be compromised. There should also be alarm systems throughout the complex that will indicate whether an intruder is trying to enter the facility grounds or a building. Video surveillance should also be considered at facilities for safety and security or where vandalism is a repeated occurrence.

Cybersecurity should be considered in the design of SCADA systems. Appropriate firewalls and internet access should be coordinated with the owner's current policies regarding restrictions. State and federal guidelines are available to provide guidance to the owner and design engineer. Guidance from the Cyber Infrastructure Security Agency or the National Institute for Standards and Technology can also be referenced.

6.2   Building Requirements

Building concepts developed during preliminary design should be finalized. Building functionality and comfort have a great effect on operator performance and morale, and the outward appearance of buildings affect community perception and acceptance. See Chapter 7 for additional discussion of architectural styles and relevant codes.

Administrative and public access areas should be designed for ADA compliance. Other areas of the facility should be designed for OSHA compliance, ensuring sufficient access for maintenance and sampling/monitoring.

An important step in building design is determining enclosed building space classification with respect to adopted building codes, and NFPA 820: *Standard for Fire Protection in Wastewater Treatment and Collection Facilities* (National Fire Protection Association, 2020), which must be coordinated with HVAC and electrical design. Fire protection may be necessary in some chemical feed and storage areas. Chemical storage areas may also require spill containment, special loading stations and/or emergency eyewash, and safety shower stations that require tempered water. Hose bibs may be necessary if washdown is common required maintenance in the area. These considerations will affect multiple disciplines involved in building design.

6.3   Power Requirements

The power requirements for new or redesigned equipment are determined by the engineer designing the equipment. Outlets should be provided at a sufficient interval for maintenance to be conducted with only one extension cord. Close coordination with the electrical engineer is required to ensure that properly sized wiring, controls, and panel gear are provided.

The need for backup power to a particular process or piece of equipment must also be determined, as this will affect the size of the backup generator or backup switchgear. Not all processes require backup power: digesters can remain unmixed or unaerated for several days without severely affecting facility performance, aerators can often be turned down, and backup equipment can be kept offline. However, some equipment requires so little power that it should be provided backup power as a convenience, regardless of whether it is absolutely needed, such as primary clarifiers and control panel user interface screens. Additional factors to consider include the type of fuel for standby power systems, difficulty of restart, effect on ability to meet permit limits, protection of environmental quality, and public health and safety impacts.

When designing backup power systems, the engineer should allow for manual or automated sequencing of equipment restart. This will help to minimize backup generator size because generator size is dependent on startup power demands, which are generally much higher than steady-state operation power demands. If equipment startup is staggered when backup power systems are operating, the size and cost of generator systems can be greatly reduced. Also, for variable-speed equipment, the engineer may limit the speed (power draw) when operating on backup power, and this may also help reduce the backup power requirement.

6.4   Heating, Ventilation, and Air Conditioning Requirements

Heating, ventilation, and air conditioning requirements at a WRRF are varied and complex. A mechanical engineer experienced in designing HVAC systems for WRRFs should be included as part of the engineer's design team. The following is a summary of WRRF spaces and goals:

- Unoccupied and occupied rooms: Keep them comfortable, dry, and odor free.

- Electrical and instrumentation rooms: Keep them cool and prohibit the entry of corrosive gases (e.g., hydrogen sulfide), which can have a detrimental effect on sensitive electrical equipment, via a slight positive pressure in the rooms.
- Process rooms: Ensure that buildup of hazardous gases (explosive, toxic, or asphyxiating gases) cannot occur, provide hazardous and explosive gas sensors, and keep the rooms from freezing. Dust control may be necessary if the facility uses dry chemicals or has a sludge drying process.

The engineer should consider area classification early in design, meeting NFPA requirements for ventilation of hazardous spaces. The placement of air intakes should also be noted, as placement too close to treatment processes may cause process off-gas or aerosols to be introduced to the HVAC system. This may be unnoticeable to facility operators if gas concentrations are sufficiently low, but long-term effects, both to human health and the longevity of sensitive electrical and mechanical equipment, can be detrimental. Also, consider freezing effects on air intakes and discharges. HVAC design must also be coordinated with the design of odor-control processes if they are to be provided as part of the facility design.

Heating, ventilation, and air conditioning systems are a common source of failed design and poor equipment selection in facility upgrades. Heating, ventilation, and air conditioning is critical to proper ventilation to ensure employee health and well-being, avoid sick building syndrome, prevent odor and corrosion protection, and improve cost efficiency of utilities.

7.0   COMPLETION OF FINAL DESIGN TASKS

Ultimately, the goal of final design is to produce a set of documents that are suitable for the solicitation of public bids for the WRRF construction project. It is important to note that there is no such thing as a "perfect design," and there will be changes during construction. See Chapter 9 for additional discussion on this topic.

Finalizing the design will constitute updating design and calculation files; finalizing specifications and drawings; and completing final reviews for safety, constructability, and operability.

7.1   Design Milestone Review Workshops

Design milestone review workshops are an opportunity to engage with the owner and facility staff to ultimately minimize design changes and build consensus on design priorities and their potential impacts.

7.1.1   Value Engineering

During the value engineering review, a team of engineers and other individuals with insight on alternative methods of achieving the project's goals will consider or propose more cost- and performance-effective design alternatives. Value engineering is generally more effective during the early stages of the design. However, if value engineering is conducted during the final design phase, it can be restricted to equipment selection, construction material choice, process and equipment layout, redundancy, bidding procedures, and operational details so that there is minimal need for design rework. For more information on value engineering, refer to Chapter 6.

7.1.2   Visualization and Operability Review

Control concepts should be reviewed for logic and ease of implementation. Valve orientation should be checked (e.g., plug valves), and the correct type of valve operator should be verified in the valve schedule or specifications (e.g., electric, lever, chain-wheel). Sampling ports and required maintenance access for equipment should be checked. If the facility has any level of automation, the ability of the entire facility to be operated in complete manual mode should be confirmed.

During the design phase of a facility upgrade, the owner should be involved in developing the format and content of the O&M manual for the engineer's scope of work. Many owners now request searchable electronic O&M manuals in addition to hard-copy documents. There are many degrees of electronic O&M manual development available, including server-based O&M manuals, which allow internet access to equipment cut sheets, parts update lists, and digital images of equipment and equipment modifications. See Chapter 10 for more information.

7.1.3   Sustainability and Resiliency Review

Sustainability and resiliency reviews should be performed during the final design to develop opportunities for reducing the project's carbon footprint and to ensure that the project is designed with the required levels of protection to function during a disruptive event or quickly come back online. Sustainability features, such as specifying recycled materials or natural lighting, can reduce project construction costs or garner public support for a project. Various levels for formal recognition of a project's sustainability are available from the U.S. Green Building Council (LEED certification) or Institute for Sustainable Infrastructure (ENVISION certification). A project's level of resiliency may be required by regulatory agencies or the project's financing program. The facility's criticality and vulnerability should be reviewed by the owner, operator, and engineer to evaluate the effectiveness of resiliency features included in the design.

7.1.4   Safety Review

Safety requirements identified in Chapter 7 should be verified during this review. Considerations should be given to minimizing confined spaces, providing safe access for routine maintenance, and color-coding or labeling piping. Examples of WRRF design safety features include fall-protection grates below wet well covers, safety shower/eyewash stations in chemical feed and storage areas, lifesaving floats for deep tankage, and stations for safety gear distribution (e.g., breathing apparatuses, earplugs, harnesses, clip-on gas alarms). Also valuable to this stage is a hazard and operability study (HAZOP) review, which involves a team with the collective experience and knowledge of the facility and an independent facilitator to rigorously challenge a design. The HAZOP identifies safety hazards and improves confidence that the design will provide a safe working environment for effective operations.

7.1.5   Constructability Review

The constructability review should identify design components that will be difficult, expensive, or awkward to construct; that may interfere with existing processes or structures; or that may not be constructable given site conditions. Such reviews should be performed by individuals with many years of experience constructing similar facilities. Comments prepared during the constructability review should be recorded and a response provided for each item, describing how the comment was addressed by the design engineer. For more information on constructability reviews, refer to Chapter 6.

7.1.6   Startup and Commissioning Review

This review should consider procedures to initiate operations of the WRRF upgrades when they are ready for testing and acceptance by the owner. The procedures for introducing flow into new facilities and steps to full operation will be identified. Special arrangements may be necessary for seeding biological treatment units, testing chemical feed systems, or maintaining permit discharge compliance when commissioning equipment. The performance requirements for the owner's acceptance should be included in the contract specifications. For more information on commissioning, refer to Chapter 10.

7.2   Quality Assurance and Quality Control

Because of the cross-disciplinary nature of facility design, QA/QC is essential to the success of a project, as poor coordination can result in conflicts and construction change orders. Each discipline must check how their work interfaces with other disciplines in addition to their typical QA/QC. Comments from these reviews should be documented and a response describing the action taken should be provided for each item.

The owner's reviewers should include individuals with experience at the facility, while the consulting engineer's reviewers should include both design team members and a specially appointed individual or group that is not involved with that particular facility design on a day-to-day basis. This will provide a "fresh set of eyes" that will catch mistakes not otherwise noted. Although it is a large time commitment, the owner's staff must carefully review the plans, drawings, and specifications to verify that all items are covered as expected.

7.3   Cost Estimates or Opinions of Probable Construction Cost

To keep track of project costs, the engineer should update the cost estimates prepared during the preliminary design phase to develop interim and final cost estimates based on equipment costs, square footage costs, material quantities and their costs, and other quantifiable costs. An interim cost estimate, often completed for the 60% submittal package, should be made to ensure that the project is not trending toward exceeding the owner's budget. Performing a cost estimate at this stage allows two possible actions: the owner can seek additional funding, or the project scope can be reduced to bring the overall costs back to the owner's original budget.

The final design cost estimate requires sufficient detail to produce a cost close to expected bid prices. The estimate should be prepared by experienced staff, and unit and equipment prices should be based on or compared to recent bidding results. The engineer's recent experience in bidding similar projects and the local bidding environment, which can be affected by labor costs, overhead costs, tax rates, and other factors, should be incorporated into the final cost estimate. The engineer's cost estimate can be used to finalize financing arrangements and can be compared to bids to determine if these are within reasonable limits—neither too high nor too low. The final cost estimate should be made just before the bid phase to ensure that the estimate represents current market conditions. The Association for the Advancement of Cost Engineering (AAEE) Recommended Practice No. 18R-97 (AAEE, 2019) provides a cost estimate classification system that is often followed when preparing cost estimates for WRRF projects.

7.4   Coordinating Bidding and Construction Procedures

Bidding procedures should be addressed in documents included at the front of the project specifications. Bidding procedures will include an invitation to bid, instructions to bidders, and the bid form itself. The instructions to bidders should state that it is the bidder's responsibility to ensure that a complete set of documents is used in determining their bid. These documents should be clear, concise, and complete.

7.4.1 The Bid Form

The bid form warrants special attention. Its format, what it includes, and what it omits will determine the type, brand, and cost of major equipment items. It can exclude undesirable equipment. It can help to minimize change orders. If it is fair, easy to understand, and complete, there will be few questions or complaints during the bidding process. A well-set-up bid form can help to determine the success of a project. The owner's legal counsel should review the bid documents.

The format of the bid form, whether it is open bid, base bid, base bid with alternates, some other type of bid, or a mix thereof, will be determined by state contracting laws and parties having jurisdiction over the project. The format may also be dictated by funding agencies.

7.4.2 Lump-Sum Allowances

Lump-sum allowances can be included in the bid form to pay for unanticipated work, such as relocation of unknown utilities or remediation of unknown hazardous materials on a time-and-materials basis. Lump-sum allowances can also be included for controls integration and SCADA work to enable the owner to maintain consistency with existing systems and adopted facility standards.

7.4.3 Unit Prices and Allowances

It may be prudent to consider the use of unit price bid items or allowances for materials or work that often result in overages. This might include adding a line in the bid form where the bidder inserts the unit price for over-excavation and refill with crushed stone. The need for this comes up often (e.g., when unanticipated buried material is encountered at an old facility, where the practice decades ago was to bury screenings on-site). By providing a generous estimate or engineer's guess of the unit quantities and having the bidder provide the unit price, the cost of changes during construction can be kept down. The actual quantities used or not used can be adjusted by change order. Unit pricing should also be used for items that cannot be precisely estimated before bidding, such as pile lengths, rock excavation, or concrete repairs.

Another use of unit prices and allowances is in requiring the contractor to buy certain equipment, preselected by the owner. In this case, it is generally prudent to negotiate a price with the selected vendor before the bid, and include that price as an allowance in the bid documents so that all bidders are working toward the same price. It should be noted, however, that preselecting specific vendors can have ramifications during construction and afterward if that equipment fails to work.

7.4.4 Open, Base, and Alternate Bid Formats

State or federal public bid requirements determine how an owner can obtain bids for the construction of a WRRF project. Conventional open bidding is the most common type of bid used in the United States. Open bids are designed to stimulate competition and require that the designer include an "or equal" clause for specified equipment.

The base bid format is also a common format used for WRRF projects and is allowed in most states. This bid format is used when an owner wants to limit the acceptable equipment that the contractor can use to prepare the bid price. State or federal approval may be required before preselecting equipment. In some cases, a separate competitive bid may satisfy this requirement. The list of preselected or preapproved equipment is prepared by the engineer and owner for inclusion in the specifications. This method allows the owner and engineer to retain control of the provided equipment but can limit competitive pricing.

This base bid with alternate bid format seeks to expand competition while still using specific criteria to maintain quality. This method requires bidders to list their price for the specified equipment, but also allows the bidder to offer a price for alternate equipment that they believe will meet the performance criteria. This bid format typically requires that the contractor name the price and specified details of the proposed alternate equipment.

It is important that bidding methods and bid forms be reviewed by the owner's attorney for compliance with the applicable bidding laws for each of the aforementioned methods.

7.5   Finalizing Construction Sequencing and Operations Plan During Construction

The project specifications should include a section describing the anticipated construction sequencing and maintenance of operations during construction. This section lists the requirements that the contractor must fulfill to ensure that components of new construction are delivered when needed and existing processes are not interrupted, or that they are interrupted based on a predetermined schedule with owner approval. It is common to define the number of process units that can be removed from service at one time, permitted electrical shutdown periods, temporary bypass or diversion plans, and who is responsible for draining and cleaning tankage and piping. A preliminary construction schedule should be included in the contract documents for complex projects to indicate suggested construction sequences or phasing, required submittal approvals, constraints, and closeout activities.

A maintenance of facility operations plan should be developed to maintain process operations and to avoid or minimize disruptions. Items that may need to be addressed include process bypasses and temporary systems for power, water, chemical feed, air, and so on.

See Chapter 9 for a complete discussion of construction sequencing.

8.0   REFERENCES

Association for the Advancement of Cost Engineering. (2019). *Cost estimate classification system as applied in engineering, procurement, and construction for the process industries* (AACE Recommended Practice No. 18R-97). AACE International.

American Institute of Architects. (n.d.). *MasterSpec.* American Institute of Architects.

British Standards Institution. (2019). *Organization and digitization of information about buildings and civil engineering works, including building information modelling (BIM)—Information management using building information modelling, Part 2: Delivery phase of the assets* (BS EN ISO 19650-2:2018). BSI Standards Ltd.

Construction Specifications Institute. (2018). *MasterFormat.* Construction Specifications Institute.

Metcalf & Eddy, Inc. (2014). *Wastewater engineering: Treatment and resource recovery* (5th ed.). McGraw-Hill.

National Fire Protection Association. (2020). *NFPA 820: Standard for fire protection in wastewater treatment and collection facilities.* National Fire Protection Association.

National Institute of Building Sciences. (2015). *National BIM Standard–United States* (NBIMS–US Version 3). National Institute of Building Sciences.

National Institute of Building Sciences. (2014). *United States National CAD Standard* (NCS –Version 6). National Institute of Building Sciences.

New England Interstate Water Pollution Control Commission. (2016). *TR-16 Guides for the Design of Wastewater Treatment Works*. New England Interstate Water Pollution Control Commission.

Water Environment Federation. (2017). *Design of water resource recovery facilities* (6th ed., Manual of Practice No. 8). Water Environment Federation.

9.0   SUGGESTED READINGS

American Water Works Association and American Society of Civil Engineers. (2012). *Water treatment plant design* (5th ed.). McGraw-Hill.

Wastewater Committee of the Great Lakes-Upper Mississippi River. (2014). *Recommended Standards for Wastewater Facilities*. Health Research, Inc.

9

Construction Phase

Arlene M. Little, PE; Brian S. Jessee, PE; & Kyle J. Blunn

1.0   CONSTRUCTION CONSIDERATIONS

The construction phase transforms the project vision into physical reality. This chapter provides insight into themes that will promote construction success. Planning for success before the start of construction is imperative. This chapter should be reviewed during the early phases of a project because many ideas require inclusion in contracts. Section 12 of this chapter on construction closeout, for instance, is an important yet commonly overlooked and underappreciated project phase. The topics in this chapter apply to any of the project delivery methods described in Chapter 5. The term "contractor," as used in this chapter, broadly refers to the entity doing the construction work.

1.1   The Unique Nature of Water Resource Recovery Facility Construction

Water resource recovery facility (WRRF) construction involves unique aspects not seen in other types of industrial or commercial construction. Wastewater from the collection system will continue to arrive during construction. The facility must remain in operation, if only partially, and must continue to meet effluent permit water quality requirements. Two parallel interrelated aspects are present in WRRF construction: the physical facilities (e.g., tanks, basins, buildings, utilities, equipment) and the processes (e.g., biological treatment, pumping, power, facility control systems). These require integrated planning and management that consider every decision during construction. Completion of the physical facilities is only part of the work.

1.2   Project-Specific Goals

Any WRRF construction project is executed to accomplish certain goals for the owner. This understanding will contribute to building the "right" facility, staying the course, and meeting the owner's drivers. Typical motivations for undertaking a WRRF construction project are discussed in Chapter 2.

1.3   Objectives for Success

Everyone on the team wants a successful construction project, but what does this mean? Typical factors for deeming a project successful or, conversely, a failure, include cost certainty, schedule certainty, quality work, safe outcomes, ease of operation, good relationships, an enjoyable experience, learning opportunities, pride, career advancements, and sense of achievement. The relative prioritization of these success factors will be unique to every project. Achieving success is measured during construction and when operating the final product. Exploring this meaning ahead of construction will better ensure a pathway to achieving project success.

Spotting construction issues early is key. The owner and engineer should act promptly and not merely trust that the contractor will or can resolve all issues. Indicators of troubled construction include schedule delays, lack of constructive communication between team members, safety violations and/or incidents, quality issues noticed by inspectors, changes in subcontractors, and lack of formal documentation.

1.4   Preconstruction Conditions

Before commencing construction, the WRRF site conditions should be established and accepted as a baseline. This can take several forms, including environmental condition reports (e.g., presence of asbestos or lead paint), photographic records, subgrade investigations (e.g., groundwater, soil types, contamination), and survey data. These must be disclosed as part of the contract. This way, work by the contractor and/or discovery of unforeseen conditions can be compared against the agreed-upon preconstruction conditions. Anything that needs to be verified by the contractor once the work commences should be clearly indicated in the contract (such as measurements that can only be taken by the contractor after a basin is drained) so that proper risk and schedule planning can be included in their work plan.

Unforeseen conditions are a common occurrence on WRRF upgrade projects. It is essentially impossible for the owner, engineer, and contractor to have complete advance knowledge about every existing feature or condition before construction. Concepts to consider to proactively mitigate negative impacts from unforeseen conditions include:

- Performing ample predesign/preconstruction site investigations (e.g., long-term groundwater monitoring, subsurface explorations)
- Recognizing shortfalls of prior project "as-built" drawings
- Identifying undocumented owner-performed construction
- Establishing a history of issues/events from long-term owner staff (e.g., spills, abandoned structures)
- Including exploratory work in the construction contract best completed by the contractor
- Allocating contingency funds

1.5   Risk Planning for the "What Ifs"

Construction, by nature, involves unknowns. "Construction is complex because uncertainty is present in almost all project activities" (Project

Management Institute, 2016). When starting construction, it is advisable to evaluate potential adverse risks that might occur; their likelihood of occurrence; and the type, extent, and magnitude of consequence. This should be performed jointly by the owner, engineer, and contractor. A suite of acceptable solutions can be considered in advance so that everyone is prepared. This evaluation should continue throughout the construction period, as an ongoing activity. This analysis will inform the amount of cost contingency the owner should carry for construction projects. Refer to Chapter 6 for further discussion on evaluating risks.

2.0   THE TEAM

People build projects. Recognizing and valuing the team is important for project success. Building a great project requires building relationships as well.

2.1   Who Is the Team?

Accomplishing construction involves a vast team. Everyone has a key role; each team member is necessary and as important as the next. Chapter 8 describes various roles during final design. Table 9.1 briefly explores additional roles and responsibilities of typical parties during construction.

2.2   Stakeholder Management

The term "stakeholder" denotes organizations or people that can affect or be affected by a project, in either a positive or negative manner. "Stakeholder satisfaction should be managed as a key project objective" (Project Management Institute, 2017). Early identification of stakeholders and ongoing management of these parties promotes project success.

A first step is to identify the stakeholders, along with their level of interest in and influence on the project execution and outcomes. Typical stakeholders for a WRRF construction project include the owner, WRRF staff, engineer, contractor, general public, local government, environmental agency, and funding agencies.

A stakeholder register should be prepared to summarize key details of the identified stakeholders and state who will be the decision-makers. Most importantly, this register should describe the respective expectations of all stakeholders, their level of interest and power, and identify the decision-makers and their influence on the project, as well as their inclinations regarding project support or resistance. During construction, the stakeholder register should be revisited and updated periodically to reflect personnel changes. For important parties, include backup staff as emergency contacts.

TABLE 9.1   Typical construction roles and responsibilities.

Role	Responsibilities
Contractor (general contractor)	Supervision, labor, materials, and equipment necessary to complete the project per contract; instill and maintain safe work environment at all times; control schedule, budget, quality control, and all project documentation
Trades subcontractors	Trade-specific expertise to construct and install to industry standards
Controls integrator	Program and develop applications needed to oversee and control facility processes during permanent operation
Suppliers	Provide specified products; generate submittals; inspect equipment installation; provide owner training
Startup and commissioning staff	Manage transition between the construction phase and permanent operation of new/upgraded equipment and processes (refer to Chapter 10)
Resident project representative	Quality assurance, oversee daily construction operations, testing, and reporting
Engineer	Review submittals, respond to requests for information, provide design adjustments, support the owner in construction contract management
Materials testing firm	Take samples and perform quality control testing

2.3   A Partnership Mindset

The team must come together in a cohesive manner. It is prudent to look beyond pure contract obligations. This involves a recognition (and self-awareness) of each member's roles, strengths, weaknesses, personality, and so forth. A mindset founded on respect should be adopted rather than an adversarial relationship. An early charter session can be conducted to collectively identify project goals and success factors. The intent is to promote open communication and foster an environment of respect and integrity throughout the project. A poster can be signed by the team to serve as a visual reminder of the partnership.

Some projects adopt a formal approach to partnering. In this case, a third-party leader facilitates the collective team's determination of construction success factors, a mission statement, and fostering team building. The duration and depth of formal partnering should be commensurate with the size, complexity, and budget of the construction project, so as to be respectful of everyone's time.

There is no project that begins and ends with the same plan and schedule. A solid team foundation will facilitate open dialogue among team members as issues arise, and collectively work toward best-outcome solutions. A relationship should never be based on a single meeting or interaction. Regular interactions will occur over the entire course of a project. The effectiveness of the entire team will be undermined if relationships break down early.

2.4   Allocation of Responsibilities and Due Diligence

The construction team, facility staff, engineer, and others will need to coexist as partners for a long duration—up to several years on substantial projects. Broad interface elements are often described in contract documents; however, these will lack sufficient detail to address daily needs.

Allocation of responsibilities should be identified in a division of responsibility (DOR) document. It is imperative to determine who is doing what, who is supplying what, and then evaluate what is falling through the cracks. Ongoing due diligence regarding allocation of responsibilities and adjusting along the way is prudent practice.

The DOR exercise will reveal contractual scope gaps. One pitfall of dividing a project into numerous contracts is that certain necessary work can become a "gap" (scope gap). These gaps are often some of the most minor work items but can have significant negative effects on the overall project success if left unattended. For instance, if the owner procured equipment for installation by the contractor, a common gap is the responsibility for equipment offloading, storing, and interim maintenance before permanent installation. A gap could also exist between the owner and engineer rather than the owner and contractor.

2.5   Staff Changes

The majority of WRRF construction projects are multi-year endeavors. It is nearly certain that there will be staff changes, whether permanent or temporary, for any of the roles. Staff may leave a project for any number of reasons including job change, retirement, internal reassignment, family emergency, geographic relocation, maternity leave, company closure, or illness. It is important to ensure that information is documented and available to the team, rather than residing with one person. Other team members should be well aware of the background for decisions and the history of events.

2.6   Contractor Qualifications

WRRF construction is a unique type of construction, involving process and timing intricacies in addition to erecting physical structures and buildings. Efficiency of execution is best accomplished by a contractor with experience in this type of work. Experienced contractors bring an in-depth awareness of WRRF-specific issues such as treatment operations, safety, bypassing, and shutdowns. Good qualifications are required not only of the contractor, but also key subcontractors in electrical, controls, and process specialties. If there are concerns regarding who should be able to propose or bid on a project, contractor prequalification is a common safeguard. Refer to Chapter 5 for contractor selection approaches.

3.0   COMMUNICATION

Construction projects are built by many people—both on- and off-site. A well-orchestrated exchange of information is an absolute must for successful construction. Many common construction problems are a result of poor communication.

3.1   Good Communication Matters

Construction entails a constant flow of information between parties, and timely responses are critical. Everyone needs something from the other stakeholders during construction, and communication allows that teamwork to be successful. Good communication means seeking a common ground of understanding, whether verbal or written, one-on-one, or in group settings. All communication should use plain language and be geared toward the audience receiving it.

Successful communication starts with a drive to achieve it. The sheer magnitude of coordination that must occur in construction is consistently underestimated by novice participants. Seek to overcommunicate early in the project to set the tone. This will ensure that all parties feel engaged and will take ownership of the project. Individuals will have preferences, strengths, and weaknesses regarding various forms of communication. Recognizing these will help facilitate effective communication. Respect and professionalism should be foundational. Behaviors such as name-calling and emotional outbursts are unproductive and undermine the goals of communication. Everyone is imperfect, and all people have days that are better than others.

The team should guard against poor information management, such as inaction on decisions, slow turnaround on reviews/responses, and conflicting directions. This often leads to construction delays and cost impacts.

3.2   Establishing Protocols

At the start of construction, the team should determine communication protocols and develop a written plan so others who enter the job at a later date know the standards. All forms of communication will be required during construction. Both verbal (speaking, listening) and written (writing, reading) communication have important purposes, as well as benefits and drawbacks. For instance, verbal conversation allows for open dialogue and relationship-building; however, these interactions are not official unless documented. Conveying important matters to multiple parties often requires using several mediums.

Communication falls along a spectrum from casual to formal. The correct mechanism should be selected depending on the situation. For example, collective brainstorming requires a facilitated yet open forum, where attendees can readily contribute to the conversation. On the other hand, certain contractual notifications require written and signed documentation delivered to the right parties.

3.3   Meetings

Construction cannot occur without meetings. Meetings allow for a common time slot for the team to efficiently exchange important information and make decisions related to construction progress. If the appropriate team members are present, meetings can lead to quick decisions that keep the project moving forward. A successful meeting starts before the meeting itself. Key recommendations include having a meeting leader, an agenda, assigned action items, and meeting notes issued in a timely manner.

A preconstruction meeting is held to launch construction as standard practice. This is often the first time all parties meet. Introductions are made and many protocols are established. Topics typically include preliminary project schedule, submittals, processing requests for payment, change orders and claims, temporary facilities, ensuring ongoing facility operations, roles and responsibilities, and safety.

A routine site team meeting held at the start of each week is recommended, attended by the contractor, facility staff, and resident project representative. Planned construction work for the week is discussed to ensure that facility staff are aware of the schedule and to allow for adjustments to the work, if necessary, depending on facility conditions.

Regular progress meetings are necessary to keep everyone informed about the project and provide an open forum for communication where everyone is present. Early in the project, select a standard day, time slot, and frequency (typically weekly or biweekly) for these recurring meetings. They are best managed by the contractor and follow a standard agenda. Topics typically include a short-term (2- to 6-week) "look ahead" schedule excerpt, safety, submittals and requests for information under review, changes, necessary coordination for upcoming work, and noncompliance or quality issues.

Planning meetings (or workshops) are recommended for coordination of upcoming specific work involving multiple parties. These are especially helpful before starting work in a new area or to plan for a temporary shutdown of the facility. This helps establish division of work, sequencing, and outstanding items that may need to be submitted, with the goal of smooth execution.

3.4   Written Documentation

Written documentation is critical for WRRF construction and helps to ensure the work will be correctly executed. Written documents constitute the records of how the construction was accomplished and what decisions were made during the work. It is easiest to address issues while "on paper" and before becoming a field issue. Examples include meeting minutes, memoranda of understanding, submittals, requests for information, and daily reports. If an event or decision is not documented in writing, it never happened. Diligence in proper documentation is imperative; documentation must be prepared in a timely manner, and must be factual, objective, unemotional, concise, organized, and dated. Written documents must reference key contract sections, include pictures if necessary, and be readily filed for retrieval at a later date. Note that some types of communication, like notification of a claim, can have a contractual deadline associated with the event.

3.5   Electronic Information Exchanges

Electronic exchange and sharing of documents and information is standard practice in modern construction. This can include shared digital platforms, email, text messages, instant messaging, and so forth. A high volume of exchanges is easily generated by using these systems. Managing, filing, and tracking these exchanges are important. WRRF construction is a long-term endeavor, and, therefore, these types of documents must be organized, filed, and easily retrievable. Many contractors use an online construction management program that can be accessed at any time by the entire team, including owners and engineers.

3.6   Conflict Resolution

Successful projects often use "soft" resolution methods such as resolution ladders or work-change meetings to maintain project momentum. In the case of a conflict between parties to a contract, a formal method of resolution is prescribed in the contract documents, such as mediation or arbitration. Maintaining positive relationships throughout the project will help prevent the need for formal resolution protocols. Refer to Chapter 10 for further discussion on problem resolution.

4.0 SAFETY

A safe job site is a successful one. Safety culture should be the focus of every stakeholder on a project from the outset.

4.1 Importance of Safety

Quality contractors and engineers have a safety mindset that starts in design and continues throughout construction, including startup and commissioning. It is not sufficient to merely follow the minimum rules. The project team has an obligation to send everyone home safe and sound after every shift. A serious injury or, unfortunately, a fatal one, will mean a job site safety shutdown. Any unanticipated site shutdowns are immediately unfavorable to the schedule and costs. Long-term negative effects to morale, productivity, and reputation can plague a project or stakeholder for many years.

4.1.1 Certifications

Certifications are required and depend on the type of work performed. Basic certifications, such as U.S. Occupational Safety and Health Administration certifications, are necessary for all contractors and their employees. Many owners require these. Certification demonstrates understanding of basic safety principles. The contractor will have designated staff (typically referred to as a "competent person") who can perform higher-level tasks such as scaffolding inspections or excavation certifications. More extensive training and certifications are needed for more serious hazards, such as working from heights, confined-space entry, asbestos abatement, arc flash protection, and other tasks.

4.1.2 Experience Modification Rate

A company's experience modification rate (EMR) is a commonly accepted "at-a-glance" indicator of a contractor's safety record. EMR is determined by insurers based on the firm's past compensation claims from workers. The number is assigned annually and used to determine insurance rates. An EMR below 1.0 is generally considered good, with lower numbers (e.g., 0.8 and below) considered top tier for safety. A higher EMR is a result of more incidents and entail higher insurance costs. The EMR is leveled across all companies based on the number of hours worked so that firms of different sizes can be directly compared by their EMR.

4.2   Hazards

Water resource recovery facility construction is considered industrial in nature because of its complexities and the juxtaposition of structures and processes. This poses certain typical industrial safety hazards such as confined spaces, heavy equipment, cranes, open excavations, high-voltage electrical work, working from heights, and operating process equipment. Workers should be trained on these hazards. During construction, protective measures are needed for workers and visitors alike, because permanent protective systems may not yet be in place.

At a WRRF, the headworks is the first arrival point of wastewater from the collection system and is often the first place wastewater encounters "open air." There is potential for release of flammable or combustible fluids and gases or low-oxygen conditions because of: (a) anaerobic conditions in the collection system that biologically generates methane and/or hydrogen sulfide and (b) potential for unauthorized discharges to the collection system (gasoline or chemicals). Similarly, within anaerobic digesters, natural biological generation of methane and hydrogen sulfide occurs. This can result in an area around the digester that has the potential for an explosive hazard. Ample ventilation, combustible gas detection, oxygen monitoring, and spark prevention should be in place during construction and before installing permanent protection systems.

Lockout/tagout is another essential safety program. Select equipment, valves, gates, and energy sources are secured closed or deenergized via a physical lock and tag to prevent accidental release of hazardous energy (e.g., hydraulic, electrical, mechanical) upon workers. Only supervisory personnel have authority to install, and more importantly, to remove locks and tags.

4.3   Interface of Contractor and Facility Safety Programs

Contractors typically control the safety program in the work area. They are required to follow governmental regulations that may exempt owners, so they must be able to control the safety program and enforce it. This often requires that an owner's personnel adopt a more stringent safety mindset during the course of the project. Conversely, the owner does have the right to enforce their safety programs (which may be more stringent) on the contractor if allowed by the contract. Many times, best practice for interfacing programs is to establish a basis of control (site-specific safety program) for the locations and activities affected by the project. As additional work areas or processes are affected, specific update meetings or trainings with the affected personnel should be conducted.

4.4   Best Practices

A safety program is most successful when the most stringent measures are adopted by all participants and are followed throughout the course of the project. Many quality contractors empower all employees, even hourly laborers, to speak out if they see unsafe acts. This is powerful, and that responsibility should also be conveyed to the owner and engineer. Other common practices include morning "toolbox talks," starting every meeting with a safety tip, or providing job-specific training to each new employee. Even discussions around near-misses can deliver a powerful message.

5.0   FACILITY OPERATIONS

Close management of interfacing with ongoing WRRF operations is necessary during any construction project.

5.1   Maintenance of Facility Operations

An existing facility must remain in reliable operation during construction, commonly termed "maintenance of plant operations" (MOPO). Construction of new elements will interface with existing systems—not just physically, but operationally—including power and controls systems. Adjoining systems require careful planning and sequencing so that existing processes continue to run reliably and new processes achieve proper startup, commissioning, and testing before being relied on to treat wastewater.

5.1.1   Process Impacts

Every facility is unique in terms of the wastewater it receives, the facility size, location, unit processes, and permit requirements. Good understanding by the construction team of how the WRRF functions is important, so construction activities can be optimally planned, scheduled, and executed. This can be accomplished by reviewing process flow diagrams, background reports (facility plans, engineering reports, regulatory agency submittals), and receiving an instructional overview from WRRF staff or the engineer. For example, a process pipe discharge should never be inadvertently connected to a storm sewer, thus creating an illicit discharge. Knowing the sensitivity of the processes is also helpful so that applicable protections can be put in place. For example, a clarifier may not be able to be taken offline if its associated chemical feed system is also offline for maintenance at the same time. Every new task that may interfere with or alter existing facilities should be reviewed and approved by the owner before proceeding.

Preparing a contingency plan for all risks is recommended and should be included in a site-specific safety plan. Its development will reveal potential problem areas to enable early resolution, before a problem becomes an emergency situation. The plan should list facility staff points of contact and on-call hours, including those with around-the-clock availability.

5.1.2 Interface of Liquid, Solids, Gaseous Systems

At a WRRF, liquid, solids, and gaseous process systems are all interrelated. Modifications to one part of the process will affect other treatment processes. It is sometimes necessary to restrict work in one area of the facility so that it does not negatively affect the performance of another area undergoing improvement. For example, full secondary treatment capacity must be ensured when biosolids dewatering modifications are occurring because of potential elevated filtrate/centrate return loading.

5.1.3 Influent Flow and Seasonal Variabilities

Understanding the inherent variability of wastewater flows coming to the facility is vital for properly scheduling construction operations. Flows and loads (extent of dilution) will vary during a 24-hour cycle (diurnal), on weekdays versus weekends, and seasonally. Local rain events will elevate flows because of inflow and infiltration into the collection system. Facility staff will have this information collected historically and should make it available to the contractor and engineer so that proper planning of shutdown or tie-in work can occur. A facility will have the greatest need for operable treatment units during times of higher flows and/or loads.

Some processes can withstand seasonal outages, while others may have to stay online nearly constantly. Seasonal variations can impart greater stress on a WRRF. In seasonal climates, secondary biological processes (aeration basins, lagoons, oxidation ditches) will treat less effectively during cold temperatures. This can limit the amount of work that can be accomplished in the winter.

5.2   Permit Compliance

It is important that the facility remains in compliance with its environmental permits throughout construction. Permits include those for treated water quality, biosolids stabilization, and air emissions. Permit violations can mean undesirable fines, additional regulatory oversight, adverse effects to the public and surrounding environment, and increased monitoring/testing. Commissioning of a new or modified process must be proven before it can be placed in service and trusted to meet permit limits or before taking another unit out of service. For biological processes, a required conditioning time may be needed before full operation is possible. It may be necessary to construct or rent a temporary unit process to allow construction to occur on an existing process being taken out of service, if sufficient redundancy does not otherwise exist. Availability of a unit process must consider the physical unit itself, as well as power feed and instrumentation and controls. New processes may require temporary connections to these support services during early construction work if a completely built-out system is not available.

5.3   Facility and Construction Interface

The contract and design documents generally focus on illustrating pre-project conditions and the intended final constructed facility and may not adequately address ongoing needs for MOPO during construction. A vast multitude of temporary facilities and arrangements are a normal part of construction. New piping, structures, equipment, and electrical systems should be built alongside existing infrastructure as much as possible. However, accomplishing final interconnections will require temporary system shutdowns, bypasses, and outages, with the goal of minimizing effects and down time to existing systems.

Any required temporary provisions known to the owner or engineer should be clearly stated in the contract so that the contractor is aware of expectations. Planning is critical.

To minimize adverse effects on one another, the construction team and WRRF staff must discuss and plan for upcoming activities together. The owner and the contractor should be committed to working as partners, so each should accept some impact to their normal operation to make the project successful for everyone. Some typical interface points are described in Table 9.2.

5.4   Constraints and the Order of Construction

The sequence of construction is affected by the expectation for utmost WRRF reliability while minimizing risks of system failure and permit violations. Good-quality designs communicate known constraints to the construction team in the contract. Constraints establish operational boundary conditions to retain reliable operations. Examples include: "At least three of four secondary clarifiers must be in service at all times" and "The new backup generator must be operational before the existing power distribution center can be upgraded." Constraints should address and consider the points of view of multiple construction trades. However, being overly prescriptive and burdensome with constraints will add to the construction schedule and costs.

TABLE 9.2. Common interface points between contractor and WRRF.

Impact	Description/Examples
Construction Impacts on WRRF	
Shutdowns, bypasses, temporary systems, outages	• Temporary system or pipe shutdown required to make critical new piping connections
	• Requirement for temporary bypass pumping, such as for a new parallel treatment basin connection or in-kind replacement of a critical pump
	• Temporary biosolids dewatering system or hauling arrangement because of replacement of biosolids dewatering and conveyance
	• Power outage required for connecting new electrical systems
Need for temporary power	• Interim power supply can be temporary cabling from existing on-site power or generators
	• Final power feed, or correct voltage, may not be available when equipment startup must occur
	• New electrical systems may be needed before old electrical equipment is replaced
	• Addition of a new backup generator to the power grid (temporary downstream power needed)
Roadway closures	• Periodic roadway closures required for installing site utility crossings
	• Temporary site construction roads are common
	• Recommend conducting discussions with local emergency service providers when roads closed or rerouted so the response time is not delayed
Site dewatering (excavations, trenches)	• Coordinate discharge point and water quality monitoring with owner and regulatory agencies
WRRF Impacts on Construction	
Site activities	• Biosolids hauling traffic, chemical deliveries, hosting construction site tours, regulatory agency inspections
Process	• Process upsets, equipment maintenance/shutdowns, abnormal influent flow/load changes from collection system

6.0 THE CONSTRUCTION CONTRACT ROADMAP

The contract is the formal legal instrument required to accomplish construction. Understanding its key elements will facilitate accomplishing successful construction. The construction contract, executed between owner and contractor, defines the mutual understanding of the work to be accomplished. Unclear or incomplete contracts will result in different understandings between the parties and often become the basis of many claims or legal disputes.

6.1   Legal Considerations

Investing in preparing a top-quality contract has many benefits. These documents outline the rules of engagement over the course of construction execution as well as define the final product. Roles and responsibilities, expectations on management style, how price negotiations are conducted, and many other standard activities are defined. Legal and insurance/risk review by knowledgeable professionals representing both parties must occur before any contract is signed. If any one party feels it is being unfairly represented, it can ask for amendments to the contract. Contractors that feel the contract is unfair and requires them to take unusual risks will price the job higher to cover their risk allocation. Refer to Chapter 5 for further discussion of these aspects.

6.2   Key Contract Elements

Refer to Chapter 8 for a discussion on types of contracts, including an overview of general project requirements, oftentimes called "Front Ends." A few contract elements are worth noting within the context of this chapter.

6.2.1 Agreement

The agreement summarizes the primary project elements, defines (and is signed by) the parties of the contract, and a date of agreement is established. It is executed following the contractor selection process (e.g., qualifications-based selection, bidding) and associated formal notification. The agreement lists associated documents that form the contract, such as drawings, specifications, amendments to the bid documents, bonds, and insurance. It includes project-specific provisions such as required milestones and payment procedures.

6.2.2 Notice to Proceed (Work Authorization)

A notice to proceed is a formal work authorization for the contractor to mobilize and commence construction. This document is typically issued after required insurance coverages are in place and the agreement has been signed.

6.2.3 Other Conditions of the Contract

Other conditions of the contract worth noting include substantial completion deadlines, work by owner or others, assignment of permitting responsibilities, contractor's responsibilities, owner's responsibilities, role of engineer during construction, payment and claims procedures, and dispute resolution mechanisms.

6.3   Protections

The contract includes risk protections for both owner and contractor. "Risk" in this context entails technical, logistical, financial, and environmental risks; these are specific to each project. It is important for both parties to identify project-specific risks before commencing construction to best mitigate their detrimental effects.

Bonds provide financial backing from a surety to ensure contract fulfillment. Performance bonds guarantee satisfactory completion of a construction contract. If the contract cannot be completed by the bonded entity (typically the contractor), the surety is responsible for completion of the remainder of the contract. This is typically completed by using a different contractor. Payment bonds guarantee that a contractor will pay its subcontractors and suppliers. If a contractor withholds payment and a subcontractor places a lien on the project, the owner can use the payment bond to pay the subcontractor and get the lien removed.

Typical types of insurance provided by the contractor include commercial general liability, automobile liability, umbrella liability, worker's compensation, builder's risk (property), and environmental hazards. Standard-form insurance certificates are prepared by the contractor's insurer(s) and supplied to the owner. The types and amounts of insurance coverage, and other related requirements, must be selected by the owner and identified in the contract.

6.4   Other Stakeholder Requirements

Other stakeholders, who are not direct parties to the contract, will have requirements that must be included in the contract. There could be work-hour restrictions or even multiple projects on the same site that will require contract provisions to be defined. Defining requirements early ensures that no obligation is forgotten and allows the contractor to include the necessary costs to comply.

6.4.1 Funding Agencies

Many WRRF projects are funded through mechanisms such as state loan funds or municipal bonds. Those agencies will stipulate requirements associated with the funding. Examples include wage rates, special purchase requirements such as American Iron and Steel and Buy American, cash-flow projection reporting, and underrepresented business participation (e.g., minority, women, disabled-veteran) requirements. These must be defined in the contract so that bidders can accurately price the work and know the terms of the contract obligations.

6.4.2   Regulatory Agencies

Regulatory agencies that hold oversight responsibilities will have project-related requirements. Examples include maintaining permit limits during the work, periodic reporting or inspections, and end-of-project submittals (overall facility O&M manual, conform to construction record drawings, certifications). Some agencies focus on the treatment process (state health or environmental departments), while others may be related to construction activities (the local stormwater pollution prevention agency).

6.4.3   Government

Some aspects of construction execution are governed by state (commonwealth, province) statutes, codes, and other legislation. Examples include non-segregation and undocumented worker legislation, retainage limits, payment timelines, licensing/permitting, and bonding. Some local jurisdictions may require certain contractor licensing/registrations, trades permitting, building permits, code inspections, and environmental requirements.

7.0   SCOPE OF THE CONSTRUCTION

The extent and nature of the construction work must be established in the contract documents. This scope of work forms the basis for the project costs and schedule and represents the baseline for comparison if extra work becomes necessary.

7.1   Defining the Work

It is important to clearly identify the scope of work to be performed by the contractor. This work includes both physical elements and services. The physical elements include new construction (e.g., buildings, basins, tanks, site work), materials, demolition, and renovations. Services include meetings, permitting, submittals, progress reporting, testing, equipment startups, temporary provisions, and owner training. In a cost-competitive environment, the contractor, subcontractor tiers, and suppliers are not inclined to include items not explicitly defined or included, as this will price themselves out of consideration. Vague scope definition will invariably result in more changes identified by the contractor. The scope is formally defined in the contract documents (refer to Chapter 8). Equally important is a clear delineation and definition of work not included (e.g., work by the owner or others, off-limit areas).

7.2   Work Breakdown Structure

A work breakdown structure (WBS) divides the work into more manageable sections for budgeting and scheduling. Each element can be further broken down to a sufficient level of detail to meet the specific needs of the project. On more complicated projects, the WBS can include constraints (outages, tie-ins, and bypasses) and selective demolition.

7.3   Project Adjustments

It is normal and expected—and even desirable—to adjust and refine the work scope, schedule, and/or cost during construction. This flexibility allows the owner to obtain the right final product and the contractor to feel that fair payment is being provided for work performed. Highly qualified contractors do not actively chase cost changes because of the added effort to prepare pricing, the resulting schedule impacts, and additional manpower needs. If the owner actively seeks to avoid cost changes "at all costs," the pressure on the team can hinder progress.

Changes can mean both additions or reductions in scope, with associated effects on the costs and/or schedule. Table 9.3 summarizes typical types of changes and examples; some are changes from what a contractor could have been reasonably expected to know at the time of baseline contract price preparation.

A common misconception is the "myth of the perfect design." The notion of a perfect design implies that the finished, future as-constructed product can be completely defined in advance, and no changes will occur to that plan during construction. In essence, the implied goal of the "perfect design" is to achieve cost and schedule certainty. While high quality is sought during all project phases, complete advanced knowledge would be akin to fortune-telling. It is only when the field team can physically dig in that many details become known. (Refer to Section 1.4 in this chapter on preconstruction conditions.)

Figure 9.1 illustrates this concept. A "perfect design" can only be approached after a long duration, meaning greater expenditure of design resources/costs to achieve smaller gains toward perfection. It is more practical to select a mutually acceptable point in time to establish the construction cost or bid the project, acknowledging that field changes are normal

TABLE 9.3   Types of construction adjustments.

Type of Change	Examples
Unforeseen conditions	Extensive groundwater, unknown buried utilities, existing structural deficiencies, changed conditions after pricing prepared
Value engineering	Alternate product or arrangement for cost savings (refer to Chapter 6)
Changes by owner	Features, facilities, work added to scope; owner-requested items not captured in design scope; convenience of having contractor readily available on-site to perform additional work
Design issue	Coordination elements, missing information. Selected equipment not performing as intended
Other parties (utility, building department, regulatory agency, etc.)	Adjustments made for code compliance, regulatory updates
Field adjustment	Small adjustments made to avoid/correct conflicts in the field
Safety, reliability, constructability improvement	Once construction takes shape, a condition can be better visualized and a preferred alternative implemented
Technological advancement	Equipment manufacturer product upgrade to be accommodated
Schedule changes	Weather delays, work by others (owner, utility company), scope addition/reduction, permitting delays, facility operational changes (e.g., loss of a piece of equipment, process upset), force majeure
Cost impacts	Extensive materials price escalations because of market factors (e.g., steel, plastics), change in supplier

practice, and can be processed in a timely manner by using a contingency or allowance.

7.4   Change Management

A change can be requested or initiated by any project stakeholder (e.g., contractor, owner, engineer) and needs to be documented in writing. The changes that will arise during construction require monitoring, processing, and tracking. The team should also understand the general tolerance for

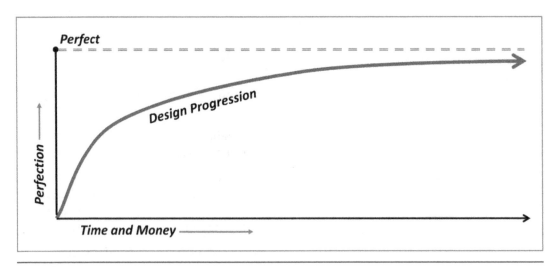

FIGURE 9.1 Myth of the perfect design.

the owner's budget to accommodate cost changes. Relatively small changes that fall below an allowable tolerance limit (cost adder or deduct) can be handled by a smaller review team, whereas larger changes must undergo more formal processes.

An established process, determined early, will benefit the project by limiting delays, expediting solutions, and minimizing conflicts. The project contract documents often describe protocols, so that the proper documentation and notifications are followed.

A designated change process group (typically the owner or engineer) should be established at the outset of construction. Sometimes this is formally called a change control board (CCB)—a group responsible for reviewing, evaluating, and approving or denying changes as well as recording the decisions made (Project Management Institute, 2017). Change control board oversight helps monitor changes compared to the baseline contract budget and schedule.

Authorizing a change requires formal written approval from the owner and/or the owner's authorized agent (e.g., the engineer). A work change directive (WCD), field order, or similar document may be issued outright to the contractor to immediately proceed with a change. Or, the owner/engineer may first want to understand resulting cost/schedule impacts before issuing a WCD and may request a preliminary pricing proposal before the work can be formalized in a WCD. A change request from the contractor is typically initiated by written notification such as a request for information (RFI) or potential change order, which could lead to WCD issuance.

When the contractor is asked to perform additional scope, a price for that work must be established. Methods for determining the cost of work include lump-sum, time and materials, and unit pricing. Early in the construction process, the owner and contractor should establish some ground rules on cost elements such as labor rates, unit pricing, escalations, and how markups can be applied (overhead and profit, bonds, insurance, materials, labor, multitier subcontractors, management time, etc.) The agreed-upon procedure should be consistent for all parties working on the project. It is common to define some procedures in the contract, so the owner should review that language during contract development to determine if it meets their financial goals. For example, the contractor may be unwilling to renegotiate markup rates after starting the work, if set rates are written in the contract.

The CCB will review the proposal and provide a written response (request to modify, approve, or deny). Timely responses are a must so that quotes remain valid and the work can progress. CCB agility is important to avoid negatively affecting construction progress via excessive formalities. The CCB should also remain objective on all changes when reviewing legitimacy; emotions should not govern whether changes are approved or denied.

8.0 THE SCHEDULE

Construction, and its multitude of tasks, has an overall duration and a necessary order of activities to accomplish the work. "Lack of planning, poor preconstruction preparation, poor communication and teamwork skills, and weak contract administration are leading causes of problems on a construction project" (Project Management Institute, 2016).

8.1 Master Schedule

A master schedule is a powerful tool. It is the overarching plan of attack required to complete the project within the contractual time limit. It is a summary of the logic and a convenient graphical means of explaining the work. A schedule assigns a duration to WBS tasks and provides relationship sequencing (e.g., excavation must take place before foundation construction). The schedule is reviewed for any constraints that may require adjustments to the duration or order of events. These could include contractual constraints (work on existing pumps cannot take place until a new pumping station is built) or environmental (exterior painting cannot take place in cold weather). When used properly, a good contractor will know when more resources are required and when to order material to ensure on-time delivery. A top-quality master schedule created by experienced professionals (and its ongoing upkeep and usage) is paramount. No successful project is accomplished without this investment. Every project finishes with an as-built schedule different from the original baseline plan. Changes are ongoing, and the master schedule requires constant attention to be most effective. Figure 9.2 is an example schedule with rollup of WBS areas.

A few select scheduling concepts are briefly described in the following subsections. In-depth discussion of scheduling principles can be found in references presented at the end of this chapter.

8.1.1 Milestones

Milestones are contractual dates that must be met. There can be a single completion project milestone, or intermediate milestones for certain systems if needed earlier. It is important to allow the contractor flexibility to implement constructability-focused solutions, because each contractor may choose to sequence the work differently to fit their needs as long as they meet the contract requirements. Flexibility in terms of the order of work, size of the labor force, material choices, alternatives with shorter lead times, and so forth can help a contractor to develop the best schedule to meet the milestones with less cost.

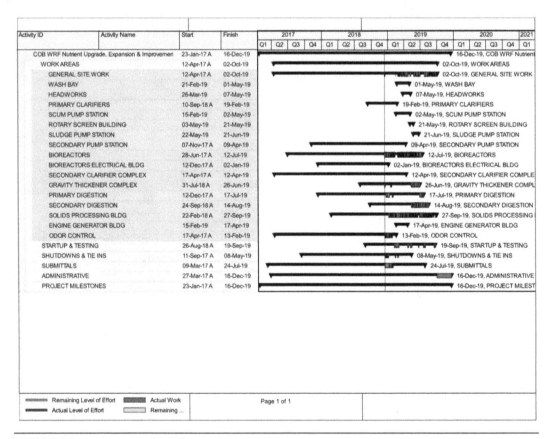

FIGURE 9.2   Example schedule.

8.1.2   Critical Path

A critical path is a series of sequential tasks that collectively form the longest overall duration of the schedule, thereby setting the completion date. If one task on the critical path is not completed on time, this creates a ripple effect on successor tasks, and the entire project is delayed. All projects have multiple, overlapping paths of the work, but only one path is critical at a time. As the project progresses, the critical path can change.

8.1.3   Float

"Float" (sometimes called "slack") is a term applied to a task indicating that a delay in accomplishing that task won't adversely affect performing other tasks. Delay can mean that the task starts later than originally intended and/ or the task requires a longer completion time. Float is essentially a "duration contingency" to ensure that overall work progress remains on schedule.

8.1.4   Cost-Loaded Schedule

In a cost-loaded schedule, a cost is allocated to each task. The net sum will equal the contract price. A cost-loaded schedule forms the basis for payments to the contractor. As items are completed and progress indicated on the schedule, the associated monetary value earned is requested on the application for payment.

8.2   Weather Impacts

Most contracts require that the baseline schedule include an allocated number of delay days caused by severe weather impacts (greater-than-average rainfall, blizzard conditions, extreme heat or cold, etc.). Adverse weather will mean the contractor cannot safely or properly perform outdoor work. Contract documents should clearly define what constitutes delays because of weather (e.g., number per month, what type of unusual weather, how to request). Typically, extra days as a result of adverse weather are added to the project duration, but the cost for lost time is not reimbursable. Schedules are not generally extended for normal weather patterns (heavy rain in spring, snow in winter, etc.) Other adverse site conditions that warrant granting a delay include residual site flooding, ice storms, high winds, or lightning storms.

8.3   Common Pitfalls

Schedules developed improperly or not updated regularly can become useless—or worse, can lead to a false sense of accomplishment. A common pitfall is lack of end planning. It is relatively easy to merely proceed upfront with a simple "four-week look ahead" and just keep updating. Unfortunately, this doesn't provide proper checkpoints to indicate if the project is falling behind, nor advance warning of long-term critical path tasks. End-of-project tasks such as startup and commissioning may be unjustly hurried to attempt to meet contractual deadlines. Deferred work may also overload site staff.

A lack of diligence in documenting schedule impacts can lead to problems. Early in the project, it is tempting to believe that everything—even small scope additions—can be "fit in" to the baseline schedule among other work. Yet, collectively over a long period, these small effects invariably lead to a missed deadline. Events that affect the schedule should be documented in a timely manner and in accordance with contract notification requirements. The owner and contractor should come to an agreement on any schedule changes at, or shortly after, the occurrence of the event. If the discussion is delayed until the end of the project, it can cause tension and must rely on memories of events that may have happened years previously. A delayed decision may have the appearance of the contractor trying to avoid contractual penalties, and/or the owner attempting to assess penalties unnecessarily.

9.0 COSTS

Understanding construction costs and good financial management principles is important. Controlling project costs is often the most recognized metric for success.

9.1 Contract Cost

The agreement between the owner and the contractor establishes the contractual monetary amount. The agreement may further break down the total contract price and may include established unit pricing, allowances, and other cost items. Understanding the baseline contract price, and associated baseline scope, is important for recognizing legitimate changes.

9.1.1 Schedule of Values

A schedule of values is a detailed breakdown of all the construction work components, with the sum equaling the contract amount. A more detailed breakdown is preferable so that progress can be tracked accurately and equitably progress payments made. The schedule of values should align with the WBS. Each item in the schedule of values should be accompanied by a fair price for that work item so that neither owner nor contractor has an advantage by accelerating or delaying payment in relation to the work completed.

9.1.2 Contingency and Allowance

Contingency and allowance are recommended for construction projects. A contingency is a fund set aside by the owner to cover the costs of unforeseen conditions, field adjustments, or owner-initiated changes. This money is typically held within the owner's financial system.

An allowance is a line item within the contract. It is used to cover unknowns that arise or certain categories for which the owner assigns a lump-sum value (i.e., the contractor doesn't prepare pricing). An allowance can be used only if approved by the owner. By drawing from an allowance, the overall cost of the project does not change. A general allowance fund can also be a depository to accept minor contract cost reductions (deducts) resulting from removed/reduced scope. This is a good mechanism to handle the ebb and flow of smaller cost changes, without requiring a formal change order for each change.

9.1.3 Change Order

A change order is executed and signed by the contract parties to officially amend the overall contract value and/or contract times. Funds for a contract increase must originate from the owner. A change order can constitute a contract increase or decrease. Refer to Section 7.4 of this chapter for a discussion on change management.

9.2 Payments to Contractor

As the work progresses, the contractor will submit for payment on work accomplished, traditionally on a monthly basis. The amount due to the contractor for an individual WBS item on the schedule of values can be based on several methodologies: percent complete, unit price, or lump sum. The collective sum is the amount due to the contractor (less any retainage).

Timely processing of pay applications and payments to the contractor is highly encouraged, and most contracts and legal statutes set limits on maximum time for payment. This supports timely payments to suppliers and subcontractors. The cooperation among the construction phase team is enhanced when monies are not unnecessarily withheld.

Retainage is an amount held back by the owner from payments to the contractor. The intent is to ensure completion of punch list items and other project closeout tasks with the incentive to be paid the retainage amount. A typical method is to withhold 5% of the amount due the contractor on each pay application. State (or provincial or commonwealth) statutes or administrative codes often prescribe the maximum allowable retainage amount and release practices. When the project has progressed to a significant degree, the owner typically releases a portion of the retainage funds to the contractor.

Ultimately, retainage is part of the contract amount, and is to be paid to the contractor with the final payment if it has not been released earlier.

9.3   Cash Flow Considerations

Cash flow is the timing of money expenditures. The contractor's cost-loaded schedule can be used to generate anticipated monthly pay application amounts over the project duration. Most owners, and their funding sources, request this information so that payment disbursements can be planned well in advance. Actual cash flow can be compared to the original forecasted plan as an independent check to assess if work is on schedule.

9.4   Using Monies to Enforce the Schedule

Construction contracts typically use financial measures to enforce the schedule. This indicates that meeting milestones is extremely important to the owner.

9.4.1   Liquidated Damages

A common financial method used is liquidated damages, which is a penalty imposed on the contractor for not meeting an important contractual milestone. Often expressed as a monetary amount per day of delay, it is intended to compensate the owner for ongoing expenses because of contractor delay. While seemingly simple in principle, liquidated damages are difficult to impose unless the delay is egregious, the owner's damages for delay are "real" and can be computed, and the definition of attaining (or missing) the milestone is clearly defined.

9.4.2   Bonus Payments

Providing a bonus payment if the contractor meets important project milestone(s) tends to strongly promote contractor efficiencies to meet the schedule in a positive manner. The entire team is encouraged to partner with the contractor to help achieve the bonus payment so that it is not a false promise. It is best to pay this bonus directly upon the contractor achieving it.

10.0   EQUIPMENT

A WRRF requires a vast amount of specialized and expensive equipment to function. The construction team must pay particular attention to the special procurement, handling, and installation requirements of equipment to be successful. Critical equipment includes major process equipment for treatment, handling, and conveyance (e.g., pumps, blowers, mixers, screens); electrical gear (e.g., motor control centers, switchgear, transformers); and control devices (programmable logic controllers, instruments). Specifics on startup and commissioning of equipment and systems are presented in Chapter 10.

10.1 Procurement

Equipment approval and procurement is a crucial part of the schedule. Contractors will determine the order in which to purchase equipment based on the project constraints and schedule. The owner may find that, to meet completion goals, equipment prepurchase may be necessary during the design stage. The engineer typically assists with this step, and the equipment is then turned over (assigned) to the contractor when it arrives. This is more common for specialized equipment with lead times of 6 months or more, or where short construction schedules do not allow for long lead times. Owner-purchased equipment should have division-of-responsibility roles clearly defined regarding storage and maintenance before installation.

10.2 Factory Testing

Major equipment will undergo testing at the factory before shipment to ensure it is functional and ready for installation. This means the equipment has received a manufacturer's quality control inspection for defects, and has been tested (mechanical, functional, hydrostatic, coating, etc.) It is more costly and time-consuming for the manufacturer to repair equipment at the site rather than at the factory where the environment is much more controlled.

Equipment factory performance testing is sometimes required in the design, such as for large process blowers and pumps. The equipment is operated under prescribed operating conditions (e.g., flow and pressure), and measurements are taken for power draw and mechanical performance at multiple setpoints. Calculations and data are produced for engineer/owner review to demonstrate that equipment performance meets the design requirements. If not, the manufacturer will modify impellers, shafts, columns, drive assemblies, or make other adjustments, and retesting is subsequently performed. These tests are sometimes witnessed by the engineer and/or owner to guard against testing bias, such as for power-intensive or critical performance equipment. It is imperative that all parties accommodate this type of witness testing in the project schedule, costs, and contracts.

10.3 Delivery, Storage, and Installation Considerations

Equipment is a significant investment. Care must be taken to protect equipment after it leaves the factory and until installation is complete.

10.3.1 Instructional Documentation

The manufacturer will prepare equipment-tailored instructional documentation to describe handling, proper storage, and installation requirements. This documentation needs to be provided to the field staff for review and understood before equipment delivery, whether to the site or to an off-site temporary storage location. This information supplements the equipment O&M manual, which focuses on long-term requirements for equipment functionality.

10.3.2 Delivery Timeline

With the specialized equipment used in WRRF processes, attention must be paid to long-lead-time items to meet the project schedule. The overall timeline is many months (4 to 6 months on average) starting with initial submittal sent to the engineer, resubmittals, engineer/owner approval, release for fabrication, manufacturing and testing time at the factory, and ending with shipment time. Some highly specialized equipment may require cross-ocean shipment. Refer to Section 12 of this chapter for a discussion of equipment warranty.

10.3.3 Storage and Interim Maintenance

Equipment should be on-site in advance of its planned installation date. This avoids unnecessary delays as a result of waiting on equipment delivery. Equipment arriving too far in advance, however, can have other risks, such as damage during storage. A high level of protection while in temporary storage will prevent unnecessary and preventable damage and repair costs. Typical storage locations include storage containers, temporary warehouses, manufacturer/supplier warehouse storage, and available WRRF storage locations. If off-site storage is used, any necessary provisions for protection, insurance, and payment should be clearly defined in the contract.

A storage plan can be useful to ensure compliance with storage and interim maintenance requirements. These requirements are typically provided to the contractor by the manufacturer. This includes both protective needs as well as interim mechanical upkeep. Examples of specific requirements are monitoring lubricant fluids, a schedule for electric motor rotation if the equipment is not installed or in use, or temperature-controlled storage of sensitive electronic equipment. The contractor is responsible for the overall project storage plan as well as quality control for all stored materials purchased, unless otherwise specified in the contract (such as for owner-procured equipment). Refer to Chapter 10 for further discussion on interim maintenance requirements.

10.3.4 Installation

Proper care during installation is important to overall performance as well as longevity of the equipment. This is also the first time that WRRF staff have laid eyes on the equipment. Staff will have valuable input on specific end-user aspects such as maintenance accessibility, drains, sampling points, and shutoff valve locations. This is an opportunity to "polish" the installation requirements to primarily benefit the operations staff. Minor modifications are more easily made before the final equipment connection.

10.4 Spare Parts

Spare parts are typically furnished with major equipment when required by the design specifications, and may include impellers, lamps, gaskets, and fuses. A spare parts plan should be developed and followed to prevent loss or damage of parts. This plan should include the chain of responsibility, allocated protective/secure storage location(s), inventory management documentation, and turnover plan and documentation. Spare parts must be delivered before startup and commissioning.

10.5 Manufacturer Site Services

After equipment is fully installed, plumbed, wired, and powered by the contractor, an authorized representative of the manufacturer will visit the site to perform startup services for major equipment. Typical tasks include installation checks, mechanical/functional checks, initial operation, on-site performance testing, and equipment training for owner staff. Certificates of completion should be prepared and signed as proof of completion for each step.

The design and subsequent purchase order for the manufacturer must include ample time and number of visits for the manufacturer's services. Work sequencing, MOPO, and staging may necessitate multiple startup trips. All parties should understand and plan for the level of effort.

11.0 QUALITY

Quality construction is a standard project success factor. The completed work should stand the test of time and be a positive testament to the project. Quality begins well before work is performed on-site.

11.1 Documentation

Several types of standard documentation allow for quality assurance. Timely reviews are imperative to ensure the momentum of construction is not broken. All required document approvals must be completed well before a product is required in the field. It is the duty of the project team to listen, review, and respond to urgent field matters in a timely manner. Some items may require accelerated reviews and responses earlier than required in the contract. The project team must plan ahead on critical submittals or other documentation requiring responses to avoid delays. Preplanning for "hot" documentation will help avoid the need for expedited reviews later. Tracking and logs are necessary to ensure that documentation is not only addressed but completed on time.

11.1.1 Request for Information

A Request for Information (RFI) is a formal means for the contractor, subcontractor, or suppliers to ask a question, seek clarity, document a decision made in a construction meeting, or raise a concern related to the construction, and obtain a documented answer. Because construction is a long process, relying on team members' memories or less official repositories (e.g., email) is a poor approach. The response to an RFI may or may not entail changes to the schedule, scope, or cost. If an RFI topic leads to such changes, it is best to follow up by issuing a WCD or change order so that cost, schedule, and overhead impacts can be properly recognized.

11.1.2 Submittals

A submittal (also commonly called a shop drawing) is prepared by a supplier, subcontractor, or the contractor to provide more precise details of a subset of the construction. Examples include equipment data, fabrication shop drawings, dimensional layouts, and material descriptions. Submittals provide a final opportunity to confirm important details and adherence to the design before ordering materials or releasing major equipment for production. The submittal also provides a record of what was installed.

In-depth reviews of submittals are a prudent investment to ensure that the proper materials and equipment arrive to the site. The contractor should perform a close review of submittals received from suppliers and subcontractors for accuracy and completeness. Submittals are reviewed by the engineer (and/or owner) who provides review comments, and the submitter will typically be asked to respond with additional information, answer questions, or resubmit the package if more extensive corrections are required. Owners are encouraged to allocate staff and time to review important equipment submittals to ensure that they are receiving the equipment inclusive of the features expected. The contractor can verify key coordination items and dimensions, but submittals require review by the engineer and/or owner to ensure compliance with the design.

Crucial coordination with other discipline trades such as electrical, controls, plumbing, mechanical, and structural work must happen during the submittal process to eliminate conflicts. Submittals provide opportunity to adjust the design if needed for coordination, constructability, changed conditions, post-design equipment manufacturer updates, and other factors. For example, the optimal placement of equipment access panels could result in adjusting pipe support locations to offer enough working clearance. If changes to the design are required based on the submittal, a WCD should be issued so that any associated project impacts (e.g., schedule, cost) are properly addressed.

11.1.3 Equipment Operations and Maintenance Manuals

An equipment O&M manual is prepared by an equipment manufacturer to provide important specifics on proper operation (normal, standby, emergency), settings, inspections, maintenance (lubrication, consumable parts), troubleshooting, and other actions. Operations and maintenance manuals should be submitted and approved as early as practical for the project and made available to the contractor installation team as well as the startup and commissioning team.

11.2   Site Inspections

Site inspections are performed by many different individuals and entities throughout the project to ensure that quality is maintained. The owner can self-perform inspections and/or hire a resident project representative, inspector, or construction manager from an outside firm to assist with inspections. Inspection personnel record observations on quality of work, answer questions about the contract documents, provide planning support, and review materials testing results.

Daily reports are paramount to recording what actually happens at a project site. These reports typically summarize daily activities, weather, workers on-site, visitors, photographs, issues, inspections, and more. These form the record of the work that occurred.

Other work inspections may be required by local building departments (e.g., plumbing, electrical), fire departments, the local authority having jurisdiction, local health department, regulatory agencies, and other entities as required by building permits, or for obtaining a building certificate of occupancy.

Special inspections may be required for specific scopes of work such as for concrete reinforcing (rebar), critical welds, soils, and deep foundations. Many local building codes allow special inspections to be performed by licensed third parties rather than by employees of the building department. This reduces the number of official inspections and allows for more flexibility in work scheduling. In most cases, expertise in WRRF construction inspection does not reside within the building department. Inspection reports should be prepared.

11.3   Quality Control Testing

Quality control testing provides objective verification of certain construction work items. If the work does not pass minimum specified requirements, then it is repaired, adjusted, or redone. A materials testing firm is traditionally hired by the owner or subcontracted through the engineer, rather than by the contractor, so that independence is maintained. Table 9.4 summarizes typical testing.

Testing for equipment and controls systems is covered in Chapter 10.

11.4   Concrete

Quality concrete is an absolute must on WRRF projects because it is broadly used for critical water-containing structures (tanks, basins, channels, digesters) and vaults, as well as for building foundations. It is quite costly to

TABLE 9.4   Typical construction quality control testing.

Item	Description/Purpose
Soil compaction	Verify that compaction and moisture content meet the minimum specified requirements to avoid settlement outside of the allowable range
Pipe leakage testing	Confirm installed pipe will hold pressure and meets the design intent for allowable leakage
Chlorine residual	Verify whether a potable service water pipe has been properly disinfected and can be placed into service
High-strength bolts	Confirm bolts have been tightened to the correct torque
Welding	Verify penetration and integrity of welds
Masonry	Confirm masonry units, mortar, and grout meet specified strength requirements
Asphalt compaction	Verify compaction of asphalt after placement to ensure strength
Drilled concrete pier placement	Confirm that no voids or unsound concrete exists by using ultrasonic logging
Basin/tank leak testing	Verify that basins or tanks (concrete, metal, or liner system) have no leaks where materials were joined

remove poor-quality finished concrete, thus close attention to quality in advance of, and during, placement is a wise investment.

Well ahead of on-site concrete placement is a period of review to finalize the concrete recipe, perform aggregate materials testing, prepare small concrete test batches, and perform testing to demonstrate that the proposed mix meets design parameters (strength in particular). This phase can be lengthy. It is imperative to allow ample time in the construction schedule for this advance work.

11.5   Design Items by Contractor Team

The contractor is sometimes required to deliver work that must be designed as part of the construction contract, commonly called a performance specification. This is distinct from a design–build contract arrangement. Examples include fire suppression/protection systems, excavation shoring systems, precast concrete, and process pipe supports. These systems must be designed by a registered professional engineer or a similarly certified and qualified professional and stamped/sealed accordingly. The inspection team can enforce the quality of the work by using the approved documents but should not take ownership.

12.0   CONSTRUCTION CLOSEOUT

To accomplish construction, contracts are executed between the owner and the contractor, between the owner and its consultants (e.g., engineer, testing agency), or others. The obligations under these contracts must be completed or reconciled for construction to be considered complete. Referencing this section while preparing those contracts is highly recommended. Contracts are notoriously early-project focused and lack adequate foresight to promote efficient completion and closeout. Specifying contract closeout procedures will create a list of actions to help guide the team at the end of the project.

12.1   Timeline

Completion of the construction phase requires a *very long* wind-down period. Planning for closeout should start well in advance. A common difficulty in efficient closeout is team fatigue. Construction veterans understand the long-term endurance and patience required to complete a project. The project team should create the closeout checklist early and strive to keep after it. Construction contracts typically describe several phases of construction completion and the associated formal documentation. Figure 9.3 summarizes these steps. The completion process is non-linear in nature with overlapping closeout activities.

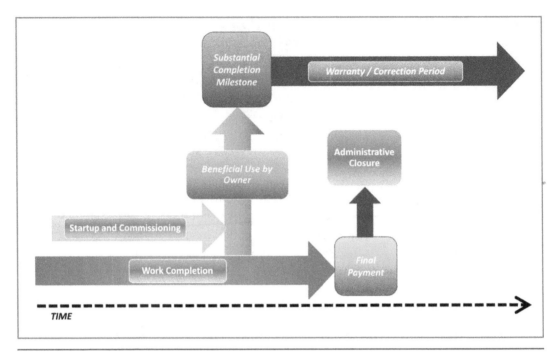

FIGURE 9.3 End-of-project timeline.

12.1.1 Substantial Completion/Beneficial Occupancy

Substantial completion is an important contractual milestone. Substantial completion is considered to be the "time at which the Work (or a specified part thereof) has progressed to the point where . . . the Work (or a specified part thereof) is sufficiently complete, in accordance with the Contract Documents, so that the Work (or a specified part thereof) can be utilized for the purposes for which it is intended" (Engineer's Joint Contract Documents Committee, 2013). This is also referred to as beneficial occupancy. Substantial completion means that all critical work has been completed, some (but not all) punch list items have been resolved, a certificate of occupancy has been obtained from the local authority having jurisdiction, and all new systems have been shown to work as designed. Typically, only minor work is left to be completed, such as punch list items, landscaping, and demobilization.

Substantial completion requirements should concisely state what work must be completed, to what extent the owner will be operating or occupying the new facilities/systems and, perhaps most importantly, what is excluded. A project can have multiple substantial completions. A formal document or certificate should be prepared to establish the date of substantial completion and be jointly signed by the owner and the contractor.

12.1.2 Punch Lists and Inspections

When the construction work has been completed to a significant degree, the contractor typically requests an inspection by the owner and/or engineer to determine if substantial completion has been met. A list of outstanding small work items is prepared, called a "punch list." A punch list can cover a specific process area or structure. All parts of the work should be completed before the punch list is created. The punch list is not intended to identify incomplete work. The punch list should include only minor items of cleanup, adjustment, and correction necessary to achieve substantial completion, final completion, and any repair of completed work that may be defective or does not meet the contract. Another inspection or inspections are conducted to confirm that the punch list work has been finished before any completion is awarded. Documented sign-offs on official inspection walk-throughs are recommended.

All logs (submittals, RFIs, action items, etc.) should be closed out by completing tasks or deferring them to another contract and documenting the associated reason. A final version of all logs should be filed and kept for records.

12.1.3 When Is Construction Complete?

Completion is achieved once all work items have been finished, including all punch list items. Construction is complete when the contractor has demobilized from the site, the owner is operating/occupying all new systems and buildings, final payment has been given, documentation provided, and all new work items are functioning to meet designed performance.

Some owners have difficulty in "letting go" of the contractor and engineer at the end of construction for various reasons, such as hesitation to fully use new systems, a habit of using the contractor and engineer to handle issues, confusion between punch list work and warranty (correction period), items and delaying the launch of expending internal costs/resources. Yet, "post-construction cost is a burden for all contracting parties" (Project Management Institute, 2016). The owner should strive to efficiently close out contracts.

12.1.4 Correction Period

The correction period, as outlined in the contract, defines the duration for which the contractor is responsible for correcting and repairing failures of the project. A project may have staggered correction periods that correlate to multiple substantial completion milestones. Be advised that many manufacturers start equipment warranty at delivery or at mechanical startup. It is the responsibility of the project team to ensure that there is clear understanding of this distinction. No warranty should be accepted as part of a submittal or as part of a delivery if it does not align with the requirements in the contract. Certain critical pieces of equipment may have extended warranty periods identified in the contract that go beyond the general project. The owner is required to perform all routine maintenance and adhere to best practices during this time because the contractor is not required to repair a failure if it was caused by owner negligence. It is common to require some insurance and bonds to remain in effect during this time to ensure performance of the repair work if needed. Refer to Chapter 10 for further discussion on equipment warranty.

12.2   Conform-to-Construction Record Drawings (As-Builts)

At construction closeout, a final conform-to-construction drawing set is prepared to reflect the work installed. The original issue-for-construction drawings are updated (either at project end and/or during the construction process) to illustrate the actual placement and dimensions of the facilities. Accurate drawings are the owner's and the engineer's most useful foundation for the next project.

12.3   Administrative Closure

Administrative closure entails confirming that contractual obligations for the contractor and the engineer have been satisfied in terms of work performed, documentation, and legal and financial obligations. Typically, a final reconciliation change order or amendment is needed for a true-up of scope, costs, and/or schedule.

Typical end-of-project documents required from the contractor include final submittal revisions or follow-up information, the as-built schedule, field plan markups and/or files from the contractor and subcontractors, equipment check-out certificates from manufacturers, warranty documents, and electronic survey files.

Closeout-phase documentation required to be submitted by the engineer typically includes conform-to-construction record drawings, regulatory agency certifications, facility-wide O&M manuals (refer to Chapter 10), and return of any reference materials belonging to the owner. Documentation from the resident project representative includes daily reports, field logs, test reports, special inspection reports, and photographic records.

A legal review is necessary to ensure all contractual obligations have been met and certificates or formal approval documents with signatures have been issued. Any remaining documentation that was in a draft version must be finalized. Any dispute resolutions need to occur and be resolved before a project is considered complete. This review ensures that no liens from subcontractors or suppliers are in place. Final insurance certificates and bonds should be submitted if any are required to be in effect after final payment is made.

Financial closure entails making final payments to the contractor and the engineer. For the construction contract, this includes releasing all retainage and paying any outstanding change orders or claims that have been approved. Many jurisdictions also require that public advertisement be made before final payment.

12.4   Other Stakeholder Agencies

Other stakeholder agencies (funding, regulatory) may have project closeout requirements such as performing final inspections and receiving construction completion (start date of operation) certifications, conform-to-construction record drawings, a facility-wide O&M manual, or other permitting documentation. Similarly, the owner may require documentation from those agencies such as certificate of occupancy, approval to discharge treated wastewater, or approval to send final payment to the contractor.

12.5   Project Reflection

Closeout is a good opportunity to reflect on the construction success factors described in Section 1 of this chapter. Were the project objectives achieved? If not, why not? Refer to Chapter 10 for a discussion of lessons learned and stakeholder satisfaction. Many team members will go on to accomplish another construction project. Taking along, and passing along, the perspectives and experiences gained will make the next project even better.

13.0   REFERENCES

Engineer's Joint Contract Documents Committee. (2013). *C-700 Standard general conditions of the construction contract.* National Society of Professional Engineers, American Council of Engineering Companies, and American Society of Civil Engineers.

Project Management Institute. (2016). *Construction extension to the PMBOK® guide.* Project Management Institute.

Project Management Institute. (2017). *A Guide to the project management body of knowledge* (PMBOK guide, 6th ed.). Project Management Institute.

14.0 SUGGESTED READINGS

Newitt, Jay S. (2009). *Construction Scheduling: Principles and Practices* (2nd ed.). Pearson Education, Inc.

Project Management Institute. (2011). *Practice standard for scheduling* (2nd ed.). Project Management Institute.

Project Management Institute. (2014). *CPM Scheduling for Construction Best Practice and Guidelines*. Project Management Institute.

10

Facility Startup and Commissioning

Georgine Grissop, PE, BCEE; EJ Katsoulas; & Jason Maskaly

1.0   INTRODUCTION

Facility startup and commissioning occur when the process units and equipment are in place and ready to be placed in operation. This applies to a new piece of equipment, upgrades to a facility, or to an entire new facility. It is the systematic process of ensuring and documenting that all components and/or systems perform interactively according to the design intent and operational needs. Figure 10.1 depicts the activities related to facility construction and commissioning. Planning and preparation are key to successful commissioning. A substantial foundation facilitates startup, performance testing, and final acceptance.

2.0   PREPARATION—STARTUP AND COMMISSIONING PLAN(S)

Startup, testing, and commissioning should first be addressed during the design phase of the project. Chapter 8 addresses startup and commissioning planning and Chapter 9 addresses maintenance of plant operations

Planning →	Design →		Bid →	Commissioning		Closeout	
Initial Facility	Preliminary	Final	Commence	Start Up	Warranty Management	Substantial Completion	Final Acceptance
Value Engineering	√	√	√				
Operability Review	√	√					
Constructibility Review		√					
Testing							
Factory			√				
Field			√				
Functional			√				
System			√	√			
Control				√			
Clean Water				√			
Performance				√			

You Are Here

FIGURE 10.1   Facility construction activities.

(MOPO). Requirements for acceptance that should be considered early in the design phase include documentation, how equipment and/or systems will be tested, and who will operate the equipment and systems and/or facility during the testing phase. If temporary connections will be needed for the testing or if waste (water, sludge, etc.) is generated during the testing, their disposal should also be considered during the design phase.

The startup phase moves from field and functional testing to more dynamic testing associated with systems and processes. During this phase, many of the following activities will take place:

- Proper operation of control systems both within a process and facility wide are verified
- Pipelines, channels, and tanks are cleared of construction debris and other foreign debris
- Piping is flushed, hydrostatically tested, and test plugs removed
- Valve alignment is double-checked
- Slide gate and stop log positions are verified
- Tanks and processes are tested with clean water
- Performance tests are run with product
- Steady-state operation is evaluated
- Training is performed

These activities cannot occur without a well-conceived startup plan prepared by a team that includes the engineer, contractor, and owner. This plan identifies team members; defines responsibilities, resources, and process strategies; and establishes sequences and schedules. The commissioning responsibilities may vary depending on the type of project delivery and should be identified in the project contract(s). Different project delivery methods are discussed in Chapter 5. Management aspects such as communication, reporting, meeting schedules, and standard agenda items should be discussed in this plan. Additionally, if the startup plan must be submitted to a regulator, it should be done early to allow for review and response to comments without affecting the construction schedule.

Goals of the startup plan include placing processes and equipment in operation in an orderly and safe manner, performing required testing to ensure that processes perform according to design, and meeting permit requirements (interim and/or final) on schedule. Responsibilities are allocated among the owner, contractor, engineer, and construction manager. The engineer who designed the process will support and/or advise on how to place systems in operation and will be consulted on process parameters to be initially used. The process parameters can be modified based on operations staff experience. The contractor's role is to ensure that all mechanical and electrical equipment are operating. A matrix should be prepared that details all activities and the party responsible for each activity. In many cases, the responsibility for an activity is shared. For example, in performance testing, the engineer, contractor, and owner may establish criteria, with the owner conducting various tests.

The startup team should be the same one that participates in construction coordination meetings, except for additional operations and maintenance (O&M) personnel. The team will include the contractor, engineer, construction manager, and representatives from process O&M.

The startup phase is intense, and many resources are required. All previous testing and construction were probably conducted during normal business hours. Now, systems are tested and operated 24 hours per day, 7 days per week. Substantial support is required from the contractor, subcontractors, vendors, operations staff, engineer, and construction manager. Resources include materials, supplies, and specialized testing equipment. Any gaps between new needs and existing resources must be filled before startup. The schedule and sequence of process startup must be well defined.

Consideration needs to be given to how unit process startup will affect ongoing operations and how interim and, ultimately, final permit obligations can be met. Although not always possible, successful clean water testing should precede the introduction of wastewater.

Ideally, before startup, all elements in Figure 10.2 are available or completed.

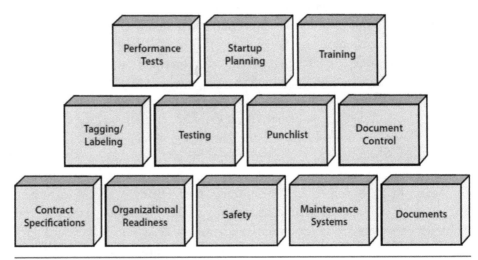

FIGURE 10.2 Prerequisites to startup.

2.1 Contract Specifications

During the design phases, the following considerations should be addressed in the contract specifications:

- Continuity of O&M during construction, including MOPO that addresses biological process, cold-weather startups, seeding, interim operations, and temporary systems (see Chapters 8 and 9 for more information)
- Detailed schedules for submission of all documents (shop drawings and vendor manuals used to create O&M manuals and facilitate training)
- Vendor manual format (print and electronic versions) and the number of final manuals
- Vendor training criteria, including examples of acceptable lesson plans, student notebooks, and specific instructor qualifications
- Number of instructor hours (class and field) and the number of training sessions (if multiple sessions are required)
- Submission and approval process for all products, manuals, notebooks, and instruction
- Maintenance information (print and electronic versions) provided in a template used by the maintenance management system
- Spare parts delivered with documentation (electronic and in print form) formatted to be compatible with warehouse and/or maintenance management systems
- Delivered equipment still under contractor care, stored and maintained according to manufacturer's recommendations
- Responsibility for maintenance of installed equipment before turnover to the owner
- Warranty information (electronic and in print form), provided in a template used by the maintenance management system
- Performance test language detailing responsibility for the supply of chemicals, power, and other required supplies

2.1.1 Operations and Maintenance Considerations

All equipment should be installed and in good working condition before commissioning. Testing may identify equipment problems that require adjustments or repairs before concluding the commissioning process.

2.1.2 Performance and Acceptance Requirements

The engineer develops performance criteria for the equipment or system and includes them in the contract documents. In addition, the engineer will confirm the product or input limits (quality and quantity), the time duration for tests, and necessary chemicals and supplies that must be provided to operations staff. Before startup, the engineer, in conjunction with operations staff, develops detailed test procedures and performance and acceptance requirements.

2.2 Organizational Readiness

A new or upgraded facility may require changes to the organizational hierarchy. Roles and responsibilities may change, new job descriptions may be required, and upgrades to a variety of operations staff certifications/credentials (operator, laboratory, collection system, crafts, or safety) may be needed. Before startup, new employees will need to be available for orientation and training activities associated with the project. The operating budget may require adjustment to provide the resources required for construction and startup. Heavy demands will be placed on the owner's staff, and unusually high overtime can be expected.

2.3 Safety Considerations

The challenge to provide a safe environment for workers intensifies during the entire construction process. Refer to Section 4 in Chapter 9 on safety.

Safety must be considered for each new process or piece of equipment undergoing startup and testing. The owner/operator should develop job safety analysis plans for the activities and provide required safety training. This includes ensuring that personal protective equipment, lockout/tagout equipment, and safety data sheets for new chemicals are available, and that eyewash showers are tested and ready.

The owner and contractor should also review heating, ventilation, and air conditioning requirements of the spaces that are to be placed in operation to ensure that area classifications are met after startup.

The project safety team must be vigilant regarding new challenges presented during the final phases of facility upgrading.

2.4 Risk Management

Before starting up and testing, the operations staff and engineer should work with the contractor personnel to help them understand the risks of starting and testing new equipment and systems. Section 1.5 of Chapter 9 addresses risk management.

2.5   Maintenance Systems and Procedures

Some facilities have a maintenance management system (MMS) that is functional during facility upgrades. The MMS is used to track work completed, parts used, and maintenance costs; assists with the planning and scheduling of work; and generates work orders (planned and corrective).

The MMS uses an equipment numbering system (asset management hierarchy) that is consistent and capable of absorbing additional information for equipment necessitated by the upgrade. Before startup, all new equipment information should be loaded to the MMS using the approved numbering system. Spare parts provided by the contractor should be identified, cataloged, and stored.

In cases where an MMS does not exist, it may be specified and procured and implemented early in the project.

2.5.1   Warranty Management

Warranty management is also a part of maintenance management. It is important to have a record of the length of time each warranty is in effect and the expiration date of that warranty. Also, the project may have had a maintenance bond associated with it. If so, this will extend the warranty, through the contractor, for the length of the bond period.

2.5.2   Interim Maintenance Requirements

As equipment arrives on-site, there may be storage, break-in, and interim maintenance requirements, such as shaft rotation to protect bearings from developing flat spots. Initial or break-in oil changes and lubricant and wear particle analyses may be required to support warranty requirements. Provisions should be made for documentation of equipment storage and interim maintenance requirements, which should then be included in the maintenance management and warranty tracking systems. In addition, initial operations documentation establishes baselines (i.e., pump curves, vibration, infrared thermography, temperature for predictive maintenance and troubleshooting). Chapter 9 covers interim maintenance from the contractor's side.

2.5.3   Spare Parts and Lubricants

As equipment and systems are placed in operation, spare parts and lubricants may be turned over to the owner. Any spare parts that have been used to make repairs during startup and commissioning will need to be replaced.

2.6   Training Program

Before startup, operations and laboratory staff should have a thorough understanding of the new processes and equipment and how they interface with existing processes and equipment. Training can be conducted in a class, on site, or off-site. In developing the training program, it is important to be sensitive to the audience because there will be different education and experience levels among personnel. Whenever possible, training should be conducted in small increments and reinforced by hands-on activities.

Equipment training covers O&M of individual pieces of equipment. The vendor or manufacturer's representative typically provides this.

Process training typically involves operation of several pieces of equipment or larger systems. Alternative modes of operation and potential problems are discussed. The engineer or an operations specialist typically provides process training.

On-site training involves the operations staff being at each unit or process and learning its operation, control, function, and maintenance.

Off-site training may be conducted at the vendor's facility if required. This is typically arranged for control or complicated equipment and where the manufacturer has dedicated training facilities.

Training takes a considerable amount of resources. Most of these resources should be addressed in the contract specifications and the scope of work developed for the engineer and construction manager. It is important that the owner include training responsibilities when developing the contracts and scope of services for the engineer. Refer to Chapter 2 for contract specifications.

Items that can be used as part of the training program may include:

- All testing (factory, field, functional, systems, and performance) results and initial settings
- Any debugging or troubleshooting reports
- Photographs and videos showing the facility at various stages of construction

At startup, a large inventory of training materials will be provided. The owner must use a document control system to catalog and store these materials. They should be readily available to operations staff during the startup and commissioning phases. Training requirement final documentation, attendance records, training hours accreditation for license renewal, and training videos should be specified.

2.7   Documents Required for Startup

Before startup, a document control system and staff to maintain it should be in place. There should be a plan to address the task of receiving, cataloging, and storing all construction and engineering documents that will be transferred to the owner. This is typically the responsibility of the construction management staff and should be part of the scope of services. Refer to Section 11 in Chapter 9 for a discussion on documentation. The general conditions of the contract specifications provide detailed guidance to the contractor and equipment suppliers concerning how, when, and in what form information is to be transmitted to the engineer and owner before startup of equipment. The following list contains most of the essential information that will be needed by the owner before startup:

- Final approved shop drawings
- Control setpoints and process control narratives
- Equipment vendor manuals
- Training submittals
- Spare parts and lubricants
- Testing results (factory, field, and functional)
- List of chemicals and supplies delivered for performance testing
- Warranties

2.7.1   Operations and Maintenance Documentation

The engineer must also provide certain information before startup, including an O&M manual, standardized operating procedures (SOPs), and overall process training to ensure that operators understand how each unit process relates to the others.

The manual should address O&M for the facility as an entire entity, with sections for each process or piece of equipment. A great deal of the information for this manual comes from the contractor and equipment supplier/vendor manuals. Chapter 9 addresses vendor manuals that build into the facility O&M. Typically, all contractor information should be submitted at the 50% level of project completion or when equipment is purchased. The O&M manual should be used as a reference for training, preparation of maintenance tasks, development of safety procedures, startup, and troubleshooting.

The manual should contain SOPs describing startup, steady-state operation, and shutdown of process equipment and systems. It should also discuss how new process units or control elements interface with existing equipment.

Sometimes, related products such as pocket guides, laminated flow-charts, and troubleshooting tables accompany the O&M manual. These products are valuable references for the operations staff and, if possible, the owner should request that these be supplied.

During the design phase, the owner should be consulted on the format and content of the O&M manual. This should be part of the engineer's scope of services. Most owners now request server-based O&M manuals. This type of manual may allow internet access to equipment cut sheets and parts update lists. It also allows for digital pictures of equipment and equipment modifications. This should be a requirement of all upgrades.

2.7.2 Checklist

This checklist includes all deliverables that should be included in the startup and commissioning plan. Table 10.1 provides examples of such deliverables.

A thorough understanding of how items are entered on the list and removed will be valuable as startup and commissioning take place. The owner should assign a staff member to review this list and ensure that all requirements are met.

2.7.3 Testing Plans

Testing plans should be submitted before startup to allow review by the engineer and owner. The plans should address the specified testing conditions, time frame, required calibrated instruments, readings to be recorded, sampling requirements, chemicals, materials, and other items as applicable to each system.

2.8   Coatings, Tagging, and Labeling

All equipment and piping should be painted according to specification and permanently tagged according to the specified numbering system (note that tags should be durable, readable, and waterproof and corrosion-proof). All valves should be tagged according to the specified tagging system, and all piping labeled according to industry standards.

2.9   Factory, Field, and Functional Testing

For large projects, hundreds of tests are performed during the construction period. Each piece of equipment is inspected for conformance with the contract documents and checked to ensure proper installation. The owner should receive a certificate of proper installation for each piece of equipment.

Preliminary information regarding test procedures should be included in the contract documents. Detailed testing procedures and forms are often

TABLE 10.1   Deliverables checklist.

Submittal schedules

Contractor's safety plan

Redline drawing updates

Vendor training:

  (1) Schedules

  (2) Training materials

  (3) Instructor qualifications

Spare parts delivery

Delivery of startup chemicals and supplies

Construction progress schedules and updates

Shop drawings

Vendor manuals

Testing:

  (1) Schedule

  (2) Results (factory, field, functional, system, and performance)

Warranties

Corrective work generated by inspections, testing activities, and changed conditions

generated with the equipment submittals and vendor manuals. Typically, these procedures are drafted by the contractor as part of the testing plan and reviewed by the engineer.

Tests cover a wide range of equipment, structures, and systems. Tests may fall into the process or nonprocess category. Examples of nonprocess testing include those performed on buildings and structures, and fire, safety, and security systems. The owner or representative, contractor, and engineer must witness these tests. The owner should request that the manufacturer's representative be present during these tests. It is recommended that operations staff be present at as many tests as possible because testing can be educational. Ideally, the same operations staff should be present at most of the tests, and then actively participate in the startup and performance phases of the project. Whenever possible, use these tests as an operations training tool. Therefore, startup and commissioning really involve completing essential building blocks (prerequisites) while performing the required work to bring new facilities into full operation and meet permit requirements.

Initially, newly completed facilities or processes rarely perform as intended. The commissioning program, which includes startup and performance testing, helps bring systems and the facility to their intended level of design. The costs of not undertaking a commissioning program can include excessively long shakedown periods during the time the facility should be performing to expectations, costly postconstruction corrections, equipment warranty problems, maintenance difficulties, and failure to meet required discharge permit requirements.

During startup, preparation ensures that influent can be introduced to the new facilities. Preparation work includes the following:

- Proper operation of control systems within a process and facility wide should be verified so that the automated systems can be used for process startup.
- Controls systems should have testing protocols specified for hardware and software.
- Once installed, control systems should be tested for each input/output point through each control sequence and as an operational unit process. This testing should include all hardwired and software interlocks.
- Distributed control systems are the "operational brains" of the facility. They are also the most challenging of facility systems to install, check out, and start. Operations staff should be included in the startup and testing of these control systems.
- Final calibration checks must be made.
- Adequate chemicals and supplies must be made available for performance testing.

After preparation work is accomplished, product is introduced to the facilities according to the startup plan. Once process conditions are established, performance testing should begin.

Activities after startup also include warranty management, substantial completion, and final acceptance.

2.10 Performance Testing

Performance testing confirms that the system operates with product and achieves the stipulated design intent. The system must meet the performance standards for a defined time period. The product input (quality and quantity) must also be within limits specified in the contract documents.

Typically, the owner conducts the testing to demonstrate that input to the system meets the contract and witnesses the formal performance test to see that the equipment meets its guaranteed performance. The engineer develops performance criteria for the equipment or system and includes these in the contract documents. The engineer will typically guide the operations staff during actual testing, interprets the results, and writes a final performance report. Delays in performance testing are common for reasons such as insufficient or incorrect chemicals, materials, and supplies, and a unit process (upstream or downstream) that is not ready or not producing the product as specified in the contract.

After performance testing is successfully completed, the facility should be able to function as designed.

2.11   Sampling and Analysis Requirements

In the interim between substantial completion and final completion, sampling is required to ensure that the treatment process is achieving the performance targets set forth in the design phase, and to aid process control and troubleshooting as needed. This is typically performed by the owner (in cases where ownership of the facility has been transferred), or in combination with the engineer and contractor. Best sampling practices should be followed to ensure the results accurately reflect the performance of the treatment process. The responsible party should ensure that they meet these six criteria for quality data:

1. Formulate objectives of the sampling plan: Establish sampling schedule, maintain consistently formatted data records, monitor results, and define a baseline for results indicating successful system performance.

2. Collect representative samples: Ensure that the correct instrument is used for each application and that it is calibrated accordingly. Collect the appropriate sample for each application (grab, time-paced composite, flow-paced composite, etc.).

3. Properly handle and preserve samples: Clean samplers, tubing, and sample containers as necessary and with the required chemicals. Do not reuse containers or other equipment that is only designed to be single use. Preserve water samples and store them in the necessary temperature conditions.

4. Properly complete chain-of-custody forms and follow sample identification procedures. If using a third party for analysis, complete all fields thoroughly to minimize confusion. Employ a consistent sample

naming philosophy and indicate all necessary details on sample containers. Apply a custody seal with all pertinent information.

5. Ensure field quality: Implement sample control (i.e., blanks, splits and duplicate samples) and equipment control (i.e., equipment maintenance logs) to maintain result accuracy.

6. Follow the proper procedures for analysis: Use an accredited laboratory that follows U.S. Environmental Protection Agency–approved testing procedures. Selected analytical methods should comply with permitting requirements and have detection limits lower than regulatory limits.

2.12   Interim Operations/Operations Transition

The interim operations and transition period responsibility should be well defined. It may be a mix of the contractor maintaining equipment with owner personnel operating the equipment. Consideration should be given for providing correct staffing/licensing for demonstration period/interim O&M. Transition to the owner may be a phased transition (i.e., first week, the contractor/engineer; second week, both contractor/engineer and owner; third week, owner with contractor/engineer available for advisement and punch list items).

3.0   PROBLEM RESOLUTION

Facility upgrading is a complicated process involving a variety of individuals and entities. Under the best of circumstances, problems will surface. As a first line of defense, problems can be eliminated by writing contracts (construction, engineering services, and construction management services) with clear expectations and realistic dispute resolution language. Further, many problems can be avoided if there is good communication (early and often) accompanied by consistent follow-up. Working as a true team can also reduce the magnitude of an issue. Making concessions instead of rigid posturing will resolve many issues without compromising the overall goals of the project. When problems do surface, the owner should solicit outside help if necessary. Chapter 9 covers conflict resolution and maintaining positive relationships with the contractor.

Often, it is helpful for the owner to speak with other users of a piece of equipment or system that is problematic to determine the best solution.

To avoid disappointment and disputes, the owner should provide input for the schedule of values review and lobby for appropriate allocation of dollars for O&M deliverables. If O&M items are late, the owner should work with the construction manager to withhold payments rather than waiting until the project is in the substantial completion stage to request deliverables.

Understanding the other parties' options for resolution is helpful in the problem-solving arena. For example, a vendor will rarely provide cash but will typically extend warranties and provide additional service and training and, ultimately, a different model or size of equipment.

There are things that the owner can do to minimize a problem or provide more time for its resolution. When possible, continue to keep existing equipment operational until all operating problems have been resolved. The owner should also devote more resources than originally called for to assist the contractor with a test or operation of contractor-supplied interim equipment.

Despite the best efforts of the team, some problems may resist resolution. In this event, follow steps outlined in the contract for dispute resolution. In all cases, good documentation is essential. All notices, costs, meeting notes, and related correspondence will be important in developing an accurate picture of the issues and framing a solution.

4.0   LESSONS LEARNED

Once startup has reached the final completion milestone, it is prudent for the owner and engineer to reflect on the successes and shortcomings of the startup and commissioning process. The formality of this assessment is at the discretion of the owner. If the desire is to produce a categorical document for reference on future collaborations, it is not uncommon for the engineer to compile a "lessons learned" report chronicling each aspect of the startup and commissioning process. For each line item, the expected and actual course of action should be listed, along with the impact of the discrepancy between the two and recommendations for improving the process. Impact categories selected should be pertinent to the project and most valued by the owner, such as cost impact or schedule impact. On the other hand, reflection and assessment can also occur as an informal meeting between the parties involved in the startup and commissioning process.

A discussion of the lessons learned report is appropriate to include during the project closeout meeting, which takes place once startup and commissioning are complete and engineering personnel are preparing to relinquish their involvement. It is important to note that reflecting on lessons learned benefits all parties involved and can foster improved relationships between these parties in future endeavors. Perhaps most importantly, taking this not-strictly-necessary step in the startup and commissioning process can streamline future projects, minimize wasted resources, and provide a significant return on investment.

5.0   COMPREHENSIVE PROJECT REVIEW AND UPDATING ORGANIZATION'S KNOWLEDGE BASE

As the project nears substantial completion, the owner should consider how to incorporate the project into the organization's knowledge base. Much of this work may have been done during the early stages of the project, such as incorporating spare parts and O&M recommendations into the owner's maintenance program. Other items may be outstanding, such as incorporating record drawings into the site-wide drawing database. The SOPs of existing systems may need to be updated to accurately reflect the iterations between existing operations and the new facility. Site permits may also need to be updated to reflect project completion. The owner should review all checklists to ensure final documentation has been received and properly filed, and no additional information is required from the engineer or vendor.

The owner should also consider how to document future changes to facility operations. Typically, SOPs are updated after substantial completion, once the facility has been operating for an extended period and operations have "settled in." Ideally, the institutional knowledge developed from the project will be documented for future reference to ensure smooth transitions between personnel and facilitate future growth and work.

If the project represents an advancement in treatment knowledge, the owner may wish to consider developing presentations and/or publications to share their expertise with others who may benefit for learning about the project.

6.0   EVALUATION OF STAKEHOLDERS' SATISFACTION

The stakeholders' satisfaction comes from the equipment or system operating and treating the wastewater to meet the facility's National Pollutant Discharge Elimination System permit. However, each stakeholder must indicate satisfaction with the new equipment or systems. Table 10.2 addresses such additional items.

TABLE 10.2 Stakeholders' satisfaction.

Owner and Engineer	Contractor	Public
System operates as designed by the manufacturer	System operates as designed by the manufacturer	New system achieves the ultimate goal of improving the well-being of the community being served
Equipment is installed correctly to ensure the full life of the equipment will be achieved	All testing is complete and documented	Equipment is aesthetically pleasing
All O&M documentation is correct and on hand		Equipment operates quietly without any alarms occurring on a regular basis
All testing is complete and documented		
All spare parts have been turned over		

Index

CPSIA information can be obtained
at www.ICGtesting.com
Printed in the USA
JSHW031152240222
23298JS00004B/68

9 781572 784055